Jekka McVicar

seeds

D0879134

Jekka McVicar

seeds

the ultimate guide to growing successfully from seed

Special photography by Marianne Majerus

THE LYONS PRESS
Guilford, Connecticut
An imprint of The Globe Pequot Press

contents

First published in Great Britain in 2001 by Kyle Cathie Limited.

First published in the United States in 2003 by The Lyons Press.

ISBN 1-58574-874-9

ALL RIGHTS RESERVED. No part of this book may be reproduced or transmitted in any form by any means, electronic or mechanical, including photocopying and recording, or by any information storage and retrieval system, except as may be expressly permitted by the 1976 Copyright Act or in writing from the publisher. Requests for permission should be addressed to The Globe Pequot Press, PO Box 480, Guilford, CT 06437.

The Lyons Press is an imprint of The Globe Pequot Press.

Text © 2001 Jekka McVicar
Photography © 2001 Marianne Majerous, except the Garden Picture Library photograph on page 88.

Senior Editor: Helen Woodhall
Editorial Assistant: Esme West
Copy Editor: Helena Attlee
Design: Vanessa Courtier
Design Assistant: Gina Hochstein
Production: Lorraine Baird and Sha Huxtable
Index: Helen Snaith

Jekka McVicar is hereby identified as the author of this work in accordance with Section 77 of the Copyright, Designs and Patents Act 1988.

Library of Congress Cataloging-in-Publication data is available on file.

Printed in Singapore by Star Standard

Hardiness Zones (as described by the US Department of Agriculture): North American gardeners are given guidance through a system of zones, of which zone 1 is the Arctic, and zone 11 the southernmost part of the continent, where temperatures never reach freezing point. A plant might be hardy to, say, zones 7-9, which means that you can leave it outdoors all winter, if you live in Texas, but not if you live in New York.

Page 2: *Cleome hassleriana* seedpods

alpines & rock plants *8*

annuals & biennials *32*

ferns *86*

grasses *92*

shrubs *172*

trees *190*

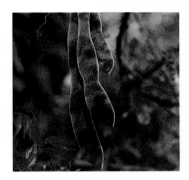

introduction

As a young child I remember being transfixed by the magic of placing a round pellet of paper into water and watching it develop into a Japanese garden. Today, I am as equally enthralled by watching tiny seeds develop into handsome plants, some of which will live for hundreds of years.

If you stop for just a moment and look closely at a single seed, you will find that it is amazing exciting, miraculous, precious and irreplaceable. Seeds come in all shapes and sizes from orchid seeds, so fine that they cannot be seen individually with the naked eye, to the largest known seed in the world—the Coco de Mer, a species of Indian Ocean palm (coconut) found on Praslin Island which is 14–16 in. in length and weighs up to 88 lb.

Each seed is a complete self-contained work of art, a unique life capsule containing the blueprint for the whole plant with every cell, hair, vein, leaf, petal and root preprogramed and waiting for germination and growth in order to manifest itself to its full potential. Seeds are masters of ingenuity when it comes to survival. Some are able to lie dormant for many seasons, or even years, waiting for the right conditions in which to germinate, others evolve slowly adapting to changes in environment. Arctic Lupin seeds have been found to be viable after lying in permafrost conditions for 10,000 years. The means of seed dispersal is equally staggering—some fly, some fall, some float. The Dandelion seed can travel 125 miles in a storm, and fern spores have been known to travel 12,400 miles from their parent. They use the air, water, animals, birds, and humans to arrive at their destination and start their cycle all over again.

The first time I sowed some seed was many years ago, when I lived in the city of Bristol. I thought it would be a good idea to be selfsufficient. So I bought lots of different packets of vegetable seed, ten seed trays, read the backs of the packets, worked out a sowing schedule, and sowed the complete seed packet into each seed tray. We had hundreds of lettuce, spinach and zucchini plants all at the same time! Undaunted by my first mistake I carried on, thinned the seedlings, pricked out the strongest seedlings and had a wonderful crop that I could share with my neighbors and my neighbors' neighbors!

From that day, the excitement of watching young seedlings appearing, as if by magic, has never left me. Even today, where in my herb nursery I grow hundreds of thousands of plants, I still get enthusiastic when I walk through the glasshouse in early spring and see the seedlings emerging. Equally, when I find a new variety, the sense of achievement that I experience when the seedling appears and then flourishes into a

new plant is very satisfying. It is this satisfaction that makes gardening such a pleasure for me and I hope that you will experience a similar sense of achievement from growing plants from seed with the help of this book.

All the plants in this book have been chosen with simple criteria in mind. They are all either common garden plants, whose seeds are simple to collect, or plants whose seeds are easily available through retailers. Turn to the suppliers' list at the back of the book if you have difficulty in finding any of them.

This book is to be used as a springboard into the world of growing plants from seed. By following a few simple steps (as simple as baking a cake), and not being too ambitious at the start, you will be rewarded by having a flourishing garden at minimal financial cost with maximum pleasure and a great sense of achievement.

alpines & rock plants

A true "alpine plant" is one that flourishes between the limits of eternal snow and the line beyond which even coniferous trees cease to grow. Some of the plants in this section fit these criteria, but others, which are also suitable for growing in rock gardens, have already adapted to milder climates. It is worth noting that quite a few of the plants mentioned in the following section do benefit from a period of cold (known as stratification) before germinating. For the best results it is, therefore, a good idea to start sowing these plants in early autumn. As these plants grow naturally in rocky conditions, it is important to make sure that they are planted in well-drained soil. Adding extra grit and sand will be of great benefit.

Viola tricolor seedhead, see page 31

Acaena *Rosaceae*

A genus of mainly low-growing perennial evergreens, which have small flowers in compact rounded heads in summer. Plant in full sun to partial shade in a well-drained soil.

Acaena microphylla AGM (Scarlet Bidi Bidi, New Zealand Bur)
Zones 7–9
Medium seeds: 2,800 per ounce
Hardy perennial
Height 3in.
Trailing habit
Bright red flowers
Flowers in summer
Gray green foliage

Sow in autumn to early spring in pots or cell packs using a standard, loam-based seed mix, mixed with 1/4 in. sharp grit. Mix to a ratio of 1 part seed mix + 1 part grit. Cover lightly with mix and place in a cold frame. Germination takes 1–3 months.

Acantholimon *Plumbaginaceae*

A genus of evergreen perennials, which have tight cushions of spiny leaves and star-shaped flowers in summer. Dislikes very wet winters, an ideal alpine house plant. If planting outside, position in full sun and in a very well-drained soil.

Acantholimon glumaceum Zones 5–9
Medium seeds: 210 per ounce
Evergreen perennial
Height 4 in.
Small spikes of star-shaped pink flowers
Flowers from late spring to early summer
Spear-shaped, spiny, dark green leaves

Sow fresh seed (germination of old seed can be erratic) in late summer in pots or cell packs, using a standard loam-based seed mix, mixed with 1/4 in. sharp grit. Mix to a ratio of 1 part seed mix + 1 part grit. Cover lightly with mix and place in a cold frame. Germination takes 1–3 months.

Adonis *Ranunculaceae*

A genus of annuals and hardy perennials, which have attractive flowers in spring and feathery leaves. Plant in a moist but well-drained soil in semishade.

Adonis annua (Pheasant's eye)
Medium seeds: 280 per ounce
Annual
Height 12 in.
Anemone-shaped, scarlet flowers which bloom singly at the tips of stems
Flowers in early spring
midgreen feathery leaves

Adonis vernalis
Medium seeds: 280 per ounce
Hardy perennial

Height 9 in.
Buttercup-like, greenish-yellow flowers which bloom singly at the tips of stems
Flowers in early spring
midgreen leaves delicately dissected

Sow fresh seed (old seed can be erratic) in late summer in pots or cell packs using a standard, loam-based seed mix mixed with 1/4 in. sharp grit. Mix to a ratio of 1 part seed mix + 1 part grit. Cover lightly with sharp grit and place in a cold frame. Germination takes 1–3 months.

Aethionema *Brassicaceae*

A genus of prolific, summer flowering, hardy, short-lived perennials. Plant in full sun and a well-drained soil. Most species self-seed.

Aethionema coridifolium (Stone cress) Zones 5–7
Small seeds: 5,600 per ounce
Hardy evergreen or semievergreen perennial
Height 6 in.
Loose sprays of tiny, rose-pink flowers
Flowers in summer
Narrow, blue-green leaves

Sow in autumn in pots or cell packs, using a standard loam-based seed mix. Cover with vermiculite or perlite and place in a cold frame. Germination takes 1–4 months.

Ajuga *Lamiaceae*

A genus of annuals and hardy perennials, the majority of which is excellent ground cover, thriving in moist conditions. Plant in partial shade or shade.

Ajuga reptans (Bugleweed) Zones 3–9
Medium seeds: 2,016 per ounce
Hardy perennial
Height 6 in.
Creeping habit
Whorls of blue flowers
Flowers in summer
Green foliage

Sow in spring in pots or cell packs, using standard soil-less seed mix either peat or a peat substitute, mixed with 1/16–1/8 in. fine grit. Mix to a ratio of 3 parts soil-less mix + 1 part fine grit. Cover with perlite or vermiculite. Place in a cold frame. Germination takes 1–2 months.

Alchemilla *Rosaceae*

A genus of hardy perennials, which have greenish-yellow flowers in summer and leaves of interesting shapes. Plant in sun or partial shade in most soils, with the exception of bogs. Most species self-seed.

Anthyllis hermanniae **Zone 8**
Medium seeds: 560 per ounce
Hardy perennial
Height 2 ft.
Small yellow, pea-like flowers
Flowers in summer
Small, bright green leaves

Sow fresh seed in autumn in pots or cell packs using standard, loam-based seed mix. Cover lightly with soil mix and place in a cold frame. Germination takes 5–6 months.
OR
Sow in spring in pots or cell packs using standard, soil-less seed mix, either peat or a peat substitute. Place under protection at 68°F. Germination takes 2–3 weeks. Flowers in its second year.

Aquilegia *Ranunculaceae*

A genus of hardy, short-lived perennials. Attractive bell-shaped, spurred flowers in spring and summer. Plant in a sunny situation and a well-drained soil. Germination of both species can be erratic, but it's worth persevering.

Aquilegia alpina (Columbine) **Zones 3–8**
Medium seeds: 1,680 per ounce
Hardy perennial
Height 18in
Clear blue or violet-blue spurred flowers
Flowers in spring and early summer
Basal rosettes of rounded, finely divided midgreen leaves

Aquilegia flabellata "Ministar" **Zones 3–9**
Medium seeds: 2,240 per ounce
Hardy perennial
Height 10 in.
Bright blue flowers with a white corolla
Flowers in summer
Rounded, finely divided midgreen leaves

Old *Aquilegia* seed is viable for five years. Sow in autumn in pots or cell packs using a standard, loam-based seed mix mixed with 1/4 in. sharp grit. Mix to a ratio of 1 part soil mix + 1 part grit. Cover lightly with soil mix, place in a cold frame. Germination takes 5–6 months.
OR
Sow fresh seed in summer in pots or cell packs, using a standard loam-based seed mix, mixed with 1/4 in. sharp grit. Mix to a ratio of 1 part soil mix + 1 part grit. Cover with perlite or vermiculite and place in a cold frame. Germination takes 3–4 weeks.

Arabis *Brassicaceae*

A genus of hardy evergreen perennials. Excellent mat-forming ground cover. Plant in a sunny situation and a well-drained soil.

Arabis alpina subsp. *caucasica* "Schneehaube" ("Snowcap"), (Rock cress, Wall cress) **Zones 4–7**
Small seeds: 8,400 per ounce
Hardy evergreen perennial
Height 5–6 in.
A profusion of single, fragrant four-petaled flowers
Flowers in spring
Oval, toothed, midgreen leaves
Excellent for a dry bank

Arabis blepharophylla "Frühlingszauber" ("Spring Charm") **Zones 4–9**
Small seeds: 5,600 per ounce
Hardy evergreen perennial
Height 5–6 in.
Fragrant carmine, single, four-petaled flowers
Flowers in spring
Oval, midgreen leaves with hairy edges
Makes excellent ground cover, but dislikes wet winters.

Sow fresh seed in pots or cell packs in early summer using standard soil-less seed mix, either peat or a peat substitute. Cover with perlite or vermiculite, place in a cold frame. Germination takes 2–3 weeks.
OR
Sow old seed in pots or cell packs in autumn using standard soil-less seed mix, either peat or a peat substitute, mixed with fine sand for extra aeration. Use a ratio of 3 parts soil-less + 1 part sand. Cover lightly with soil-less mix and place in a cold frame. Germination takes 3–4 months.

Areneria *Caryophyllaceae*

A genus of annuals and hardy perennials which flower in the spring and summer. Plant in a sunny situation in a well-drained, preferably sandy, soil.

Areneria montana (Sandwort, Mountain sandwort) **Zones 3–8**
Medium seeds: 2,380 per ounce
Hardy perennial
Height 2 in.
Large white, round flowers
Flowers in summer
Loose mats of small, narrow, oval, midgreen leaves

Sow in spring in pots or cell packs using standard, soil-less seed mix, either peat or a peat substitute. Cover with perlite or vermiculite and place under protection at 68°F. Germination takes 4–6 weeks. Flowers in its second year.

Armeria *Plumbaginaceae*

A genus of evergreen perennials which grow in tufty clumps and flower in the spring and summer. Plant in a sunny situation in a well-drained soil.

Armeria maritima (Sea pink, Thrift) **Zones 3–8**
Medium seeds: 1,120–1,960 per ounce
Hardy evergreen perennial
Height 4 in.

Small pink or white, globe-shaped flowers
Flowers in early summer
Leaves are dark green and grass-like, growing in little mounds
Good as an edging plant

Sow in autumn in pots or cell packs using a standard, loam-based seed mix. Home-collected seed will need to be scarified (see page 233). Seed is viable for 3 years. Cover lightly with seed mix. Place in a cold frame. Germination takes 2–3 months, but is erratic.

Aubrieta *Brassicaceae*

A genus of hardy evergreen perennials, which grow in mounds or are trailing. Flowers in the spring or summer. Plant in a sunny situation and a well-drained soil.

Aubrieta pinardii "Red Cascade" Zones 4–8
Small seeds: 7,000 per ounce
Hardy evergreen perennial
Height 4 in.
A profusion of carmine-red flowers
Flowers in early summer
Rounded, toothed, soft midgreen leaves

Sow in spring in pots or cell packs using standard, soil-less seed mix, either peat or a peat substitute. Cover with perlite or vermiculite and place in a cold frame or under protection at 60°F. Germination takes 2–3 weeks cold ,or 7–14 days with protection. Flowers during the following year.

Aurinia *Brassicaceae*

A genus of hardy evergreen perennials which either grow in mounds or are trailing. Flowers in the spring or summer. Plant in a sunny situation in a well-drained soil

Aurinia saxatilis "Gold Dust" Zones 3–7
Medium seeds: 2,800 per ounce
Hardy evergreen perennial
Height 6 in.
Small racemes of four-petaled yellow flowers
Flowers in late spring
Small, oval, hairy, gray green leaves

Sow in spring in pots or cell packs using standard, soil-less seed mix, either peat or a peat substitute. Cover with perlite or vermiculite and place in a cold frame or under protection at 68°F. Germination takes 2–3 weeks cold, or 3–8 days with protection. Flowers during the following year. Excellent for growing over a wall.

Ballota *Lamiaceae*

A genus of frost-hardy perennials and evergreens. Flowers in the summer. Plant in a sunny situation and a well-drained soil.

Ballota pseudodictamnus AGM Zones 7–10
Medium seeds: 1,960 per ounce
Hardy, evergreen perennial
Height 24 in.

Whorls of pink flowers with pale green calyces
Flowers in summer
Hairy/woolly, gray green, rounded leaves
Good, large rock garden plant with all year interest

Sow in autumn in pots or cell packs using a standard loam-based seed mix. Cover lightly with soil mix and place in a cold frame. Germination, which is erratic, takes 2–4 months.

Bellis *Asteraceae*

A genus of perennials, cultivars often grown as biennials. Plant in sun to semishade in a fertile soil that is well-drained.

Bellis perennis English daisy) Zones 3–8
Small seeds: 13,720 per ounce
Hardy perennial
Height 6 in.
Small, flat, double-petaled, white flowers with a touch of pink
Flowers in early summer
Oval, midgreen leaves

Sow in summer in pots or cell packs using standard, soil-less seed mix, either peat or a peat substitute. Cover with perlite or vermiculite and place in a cold frame. Germination takes 2–3 weeks. Flowers the following season

Calandrinia *Portulacaceae*

A genus of perennial and annual plants, which have a spreading or trailing habit and fleshy leaves. The flowers close in dull weather and at night. Plant in a sunny situation and well-drained soil.

Calandrinia umbellata (Rock purslane)
Small seeds: 25,700 per ounce
Perennial, grown as an annual
Height 4–6 in.
Vivid crimson, cup-shaped flowers
Flowers in summer
Gray green, slightly hairy leaves

Sow in spring in pots or cell packs, using standard soil-less seed mix, either peat or a peat substitute. As these are very fine seeds, mix them with the finest sand or talcum powder for an even sowing. Cover with perlite or vermiculite, place in a cold frame. Germination takes 3–4 weeks. Might flower in its first season.

Campanula *Campanulaceae*

A genus of hardy perennials, biennials, and annuals which flower in the spring and summer. Plant in sun or partial shade in a moist but well-drained soil.

Campanula carpatica AGM (Carpathian harebell) Zones 3–8
Small seeds: 19,600 per ounce
Hardy perennial
Height 4 in.
Broadly bell-shaped, blue or white flowers
Flowers in summer
Rounded to oval, toothed green leaves

Sow in spring in pots or cell packs using standard, soil-less seed compost (substrate), either peat or a peat substitute. Cover with perlite or vermiculite. Place in a cold frame, or under protection at 68°F. Germination takes 3–4 weeks, or 1–2 weeks with heat. Can flower in first season in midsummer.

Carlina *Asteracea*

A genus of annuals, biennials, and perennials. All have attractive flowers. Plant in a sunny situation in a well-drained soil.

Carlina acaulis subsp. *simplex* (Silver thistle) Zones 4–7
Medium seeds: 700 per ounce
Hardy perennial
Height 6 in.
Stemless, single, off-white or pale brown flowers with papery bracts
Flowers in summer
Long, deeply cut leaves with a spiny margin, forming rosettes

Sow in autumn in pots or cell packs, using a standard loam-based seed mix. Cover lightly with seed mix and place in a cold frame. Germination takes 4–6 months. Flowers the following season.

Cerastium *Caryophyllaceae*

A genus of annuals and perennials with star-shaped flowers. Plant in a sunny situation in a free draining soil.

Cerastium tomentosum (Snow in summer) Zones 4–7
Small seeds: 7,000 per ounce
Hardy perennial
Height 3 in. creeping habit
Star-shaped, pure white flowers
Flowers in summer
Small, silvery-gray, serrated leaves

Sow in autumn in pots or cell packs using a standard, soil-less seed mix, either peat or a peat substitute, mixed with fine sand for extra aeration. Mix to a ratio of 3 parts soil-less mix + 1 part sand. Cover with perlite or vermiculite and place in a cold frame. Alternatively, sow in spring and place under protection at 65°F. Germination takes 2–4 months cold, or 1–2 weeks with warmth. Can flower in its first season. Makes a vigorous ground-cover plant.

Chaenorhinum *Scrophulariaceae*

A genus of hardy annuals, biennials, and perennials. Plant in sun or light shade in a well-drained soil.

Chaenorhinum originafolium "Blue Dream"
Small seeds: 3,500 per ounce
Hardy perennial, grown as an annual
Height 5 in.
Cushion habit
Lots of small, pale blue or mauve flowers
Flowers in summer
Small, oval midgreen leaves

Sow in spring in pots or cell packs using standard soil-less seed mix, either peat or a peat substitute. Cover with perlite or vermiculite and place under protection at 65–68°F. Germination takes 1–2 weeks.

Codonopsis *Campanulaceae*

A genus of hardy perennials and herbaceous, twining climbers. Plant in partial shade in light, well-drained soil.

Codonopsis clematidea Zones 5–9
Small seeds: 4,450 per ounce
Hardy perennial
Height 20 in.
Light blue, bell-shaped flowers with a white tinge
Flowers in summer
Ideal for banks and dry walls

Sow in spring in pots or cell packs using standard, soil-less seed mix, either peat or a peat substitute. Cover with perlite or vermiculite and place in a cold frame or under protection at 65–68°F. Germination takes 1–6 weeks. or 1–3 weeks with warmth.

Corydalis *Papaveraceae*

A genus of hardy annuals, perennials, and evergreens some with tuberous or fibrous roots. Plant in partial shade or full shade in a well-drained soil.

Corydalis lutea (Golden bleeding heart) Zones 5–7
Medium seeds: 2,100 per ounce
Hardy evergreen perennial
Height 12 in.
Yellow flowers with short spurs, which grow in dense racemes
Flowers from spring and through the summer
Basal, divided, gray-green leaves

Sow in autumn or early spring in pots or cell packs, using a standard loam-based seed mix mixed with coarse horticultural sand. Mix to a ratio of 1 part soil mix + 1 part sand. Cover lightly with mix and place in a cold frame. Germination takes 3–12 months (do not give up). Flowers during the following year.

Cyclamen *Primulaceae*

A genus of hardy to frost-tender tuberous perennials, some evergreen, all with attractive pendent flowers. Plant in a humus-rich, well-drained soil in partial shade.

Cyclamen hederifolium (syn. *Cyclamen neopolitanum*) Zones 5–9
Medium seeds: 280 per ounce
Hardy perennial
Height 6 in.
Lovely, small, pink, pendant flowers
Flowers from late summer to early autumn
Deep green, ivy-shaped leaves with attractive silver markings

Soak fresh seeds in warm water with a little liquid soap for 12 hours prior to sowing. Sow in autumn in pots using standard, loam-based seed mix, mixed with coarse horticultural grit. Mix to a ratio of 1 part soil mix + 1 part 1/4 in. grit. Cover with sharp grit. Place the container under protection at a minimum temperature of 60°F. Germination takes 3–4 weeks. Flowers during its second season.

Recipe for sowing, planting, and growing Cyclamen hederifolium

This is a charming plant. It looks lovely grown in clusters or drifts on a rocky bank or under deciduous trees. The seeds of *Cyclamen hederifolium* take a season to ripen on the plant. In nature, the seed capsules coil down to the ground. They are covered in a sticky coating which is attractive to ants—Nature's seed sowers of this species.

Ingredients
2 seeds per cell or 4 seeds per pot
Small bowl
Warm water with a minute dash of liquid soap
1 x 4 in. pot
Standard loam-based seed mix
Coarse horticultural grit
Additional coarse grit for covering
White plastic plant label

Method Collect fresh seeds in autumn. Fill a bowl with warm water. Add a minute dash of liquid soap. Soak the fresh seeds in the liquid for 12 hours. Fill the pot with a mixture of soil mix and grit, mixed to a ratio of 1 part soil mix + 1 part grit. Smooth over, tap down and water in well. Put four seeds in each pot, spaced equally and placed on the surface of the soil mix. Cover the seeds with sharp grit and label with the plant name and date. Place the pot in a warm spot out of direct sunlight at an optimum temperature of 60°F.

Keep watering to a minimum until germination has occured, which takes 3–4 weeks with warmth. Shortly after germination, when the seedlings have emerged fully, put the container into a cold frame for the winter. If you have not overcrowded the pot with seedlings it is a good idea to leave them in this pot undisturbed for the season. Split the tubers when they are dormant, repot in the same soil mix and put the young plants in the cold frame for the rest of the winter. Plant out into a prepared site in the garden in spring.

Dianthus *Caryophllaceae*

A genus of annuals, biennials, perennials, evergreens or semi-evergreen plants grown for their flowers. They are often scented and some are excellent for cutting. Flowering is mainly from late spring to late summer. The majority prefers to be planted in full sun and in a well-drained, slightly alkaline soil.

Dianthus deltoides AGM (Maiden pinks) Zones 3–9
Small seeds: 4,760 per ounce
Hardy, evergreen perennial
Height 6 in.
Mat-forming
Small white, pink or cerise flowers
Flowers in early summer
Narrow, lance-shaped, dark green leaves

Dianthus gratianopolitanus (caesius) AGM (Cheddar pink)
Zones 4–9
Medium seeds: 1,960 per ounce
Hardy, evergreen perennial
Height 8 in., creeping habit
Pale pink, very fragrant flowers in early summer
Narrow, gray green leaves

Sow in spring in pots or cell packs using standard, soil-less seed mix, either peat or a peat substitute. Cover with perlite or vermiculite and place in a cold frame or under protection at 68–72°F. Germination takes 2–3 weeks, or 7–10 days with warmth.

Draba *Brassicaceae*

A genus of annuals and evergreen or semievergreen perennials. Plant in a gritty, well-drained soil in a sunny situation. Hates wet winters.

Draba aizoides (Yellow whitlow grass) Zones 4–8
Small seeds: 9,800 per ounce
Hardy, semievergreen perennial
Height 1 in.
Bright yellow, four-petaled flowers
Flowers in spring
Lanced-shaped, bristle-like, gray green leaves

Sow in spring in pots or cell packs using standard, soil-less seed mix, either peat or a peat substitute. Cover with perlite or vermiculite, place in a cold frame. Germination takes 2–3 weeks.

Dryas *Rosaceae*

A genus of evergreen, prostrate perennials. Plant in a gritty, peaty, well-drained soil in a sunny situation.

Dryas octopetala AGM (Mountain avens) Zones 2–7
Medium seeds: 2,800 per ounce
Hardy, evergreen perennial
Height 2 in., prostrate, mat-forming habit
Cup-shaped, creamy-white flowers
Flowers in spring and early summer
Leathery, oval, dark green leaves

Sow with fresh seed in autumn in pots or cell packs using standard, soil-less seed mix, either peat or a peat substitute, mixed with 1/16–1/8 in. fine grit. Mix in a ratio of 3 parts soil-less mix + 1 part fine grit. Cover lightly with mix and place in a cold frame. Germination takes 2–3 months.

Erinus *Scrophulariaceae*

A genus of semievergreen, short-lived perennials. Plant in a well-drained soil in a sunny situation.

Erinus alpinus AGM (Lever balsam, Fairy foxglove) Zones 4–9
Tiny seeds: 44,800 per ounce
Hardy perennial
Height 3 in.

Herniaria *Caryophyllaceae*

A genus of hardy, short-lived, evergreen perennials. Plant in well-drained soil in sun or semishade.

Herniaria glabra (Rupture wort) Zones 5–9
Small seeds: 14,000 per ounce
Hardy, evergreen perennial
Height 4 in.
Creeping habit
Small, green flowers
Flowers in summer
Small, oblong, deep green, grass-like leaves

Sow in spring in pots or cell packs using standard, soil-less seed mix, either peat or a peat substitute. Cover with perlite or vermiculite, place in a cold frame, or under protection at 65–70°F. Germination takes 2–3 weeks cold, or 5–15 days with warmth. Flowers during its second season. Good for growing in walls.

Heuchera *Saxifragaceae*

A genus of hardy evergreen perennials. Excellent for ground cover. Plant in moist but well-drained soil in semishade.

Heuchera micrantha var. *diversifolia* "Palace Purple" AGM **Zones 4–8**
Tiny seeds: 56,000 per ounce
Hardy perennial
Height 18 in.
Sprays of small white flowers
Flowers in summer
Heart-shaped, deep purple leaves

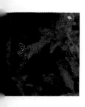

Sow in spring in pots or cell packs using standard, soil-less seed mix, either peat or a peat substitute. As these are very fine seeds, mix them with the finest sand or talcum powder for an even sowing. Do not cover. Water from the bottom or with a fine spray and place in a cold frame or under protection at 70°F. Germination takes three weeks cold, or 5–15 days with warmth. Flowers during its second season.

Hieracium *Asteraceae*

A genus of hardy perennials. Plant in a well-drained soil in a sunny situation.

Hieracium villosum Zones 5–8
Medium seeds: 1,680 per ounce
Hardy perennial
Height 6 in.
Bright yellow, dandelion-like flowers
Flowers in summer
Silvery-gray, woolly leaves

Sow in spring in pots or cell packs using standard, soil-less seed mix, either peat or a peat substitute. Cover with perlite or vermiculite and place in a cold frame or under protection at 69°F. Germination takes 2–3 weeks cold, or 5–15 days with warmth. Flowers during its second season.

Hypericum *Clusiaceae*

A genus of yellow, flowering perennials, semievergreens, evergreens, shrubs, and subshrubs.

Hypericum calycinum Zones 6–9
Small seeds: 6,700 per ounce
Evergreen or semievergreen dwarf shrub
Height 12 in.
Bright yellow flowers
Flowers from midsummer to midautumn
Dark green, oval leaves

Hypericum polyphyllum
Small seeds: 9,520 per ounce
Hardy, evergreen perennial
Height 8 in.
Bright yellow flowers all summer
Small, finely textured, midgreen leaves

Sow from early to midspring in pots or cell packs using standard, soil-less seed mix, either peat or a peat substitute. Cover with perlite or vermiculite and place in a cold frame or under protection at 65°F. Germination takes 1–3 months cold, or 10–21 days with warmth.

Iberis *Brassicaceae*

A genus of annuals perennials, evergreens, and shrubs. All are very good rock garden plants. Plant in a well-drained soil and a sunny position.

Iberis sempervirens (Candytuft) Zones 3–9
Medium seeds: 924 per ounce
Hardy, evergreen subshrub
Height up to 12 in.
Dense clusters of round heads of white flowers
Flowers from spring to early summer
Narrow, oblong, dark green leaves

Sow in spring in pots or cell packs using standard, soil-less seed mix, either peat or a peat substitute. Cover with perlite or vermiculite and place in a cold frame or under protection at 65°F. Germination takes three weeks cold, or 14–21 days with warmth. Flowers during its second season.

Leontopodium *Asteraceae*

A genus of short-lived hardy perennials grown for their attractive flowers. Plant in a well-drained soil in a sunny position. These plants hate being overly wet at any time of year, so plant with a good collar of sharp grit.

Leontopodium alpinum (Edelweiss) Zones 4–7
Small seeds: 21,000 per ounce
Hardy perennial (short-lived)
Height 6–8 in.
Lovely, silvery-white flowers surrounded by star-shaped bracts
Flowers in spring
Lance-shaped, slightly woolly, silvery-green leaves

Iberis sempervirens

Sow in spring in pots or cell packs using a standard, loam-based seed mix, mixed with 1/4 in, sharp grit. Mix to a ratio of 1 part soil mix + 1 part grit. As these are very fine seeds, mix them with the finest sand or talcum powder for an even sowing. Do not cover. Water from the bottom or with a fine spray, place in a cold frame, or under protection at 65°F. Germination takes three weeks cold or 14–21 days with warmth. Flowers in its second season.

Lewisia *Portulacaceae*

A genus of hardy perennials and evergreens. Plant in a rich, moist or well-drained, neutral to acid soil in semishade. Herbaceous species should be planted in full sun.

Lewisia cotyledon AGM **Zones 5–9**
Small seeds: 4,200 per ounce
Hardy evergreen perennial
Height up to 12 in.

Clusters of small flowers in shades of pink
Flowers in summer
Large, thick, midgreen leaves
Hates being wet in winter, so put a collar of fine grit around it
Ideal for rock gardens and alpine houses

Lewisia nevadensis **Zones 6–9**
Small seeds: 4,200 per ounce
Hardy perennial
Height 1–2 in.
Pretty, cup-shaped, white flowers
Flowers in summer
Narrow, dark green leaves

Sow in spring in pots or cell packs using standard soil-less seed mix, either peat or a peat substitute. Cover with perlite or vermiculite and place in a cold frame or under protection at 65°F. Germination takes 1–2 months cold, 14–21 days warm. Flowers in its second season.

Linaria *Scrophulariaceae*

A genus of annuals, biennials, and perennials. Plant in any well-drained soil in sun or partial shade.

Linaria alpina (Alpine toadflax) Zones 4–9
Small seeds: 12,600 per ounce
Hardy perennial (short lived)
Height 6 in.
Purple and violet flowers with yellow centers that grow in loose racemes
Flowers in summer
Linear to lance-shaped, fleshy, gray green leaves

Linaria cymbalaria (Cymbalaria muralis) (Ivy-leaved toadflax)
Zones 4–9
Small seeds: 14,000 per ounce
Hardy perennial

Height 2 in.
Creeping habit
Small, snapdragon-like flowers in purple, mauve, and white
Flowers in summer
Bright green, round lobed leaves
Lovely growing in walls

Sow in spring in pots or modules using standard, soil-less seed mix. either peat or a peat substitute. Cover with perlite or vermiculite and place in a cold frame. Germination takes 2–3 weeks. Flowers in the first year.

Linum *Linaceae*

A genus of annuals, biennials, perennials, and evergreens grown for their attractive flowers. Plant in a rich, but well-drained, peaty soil in a sunny situation.

Linum perenne

Linum flavum (Golden flax, Yellow flax) Zones 5–8
Medium seeds: 2,240 per ounce
Hardy perennial
Height 12 in.
Clusters of yellow, funnel-shaped flowers
Flowers in summer
Narrow, oval, green leaves

Linum perenne Zones 4–9
Medium seeds: 840 per ounce
Hardy perennial
Height 12 in.
Clusters of open, funnel-shaped, lovely, clear blue flowers
Flowers in summer
Grass-like, slender leaves

Sow in spring in pots or cell packs using standard, soil-less seed mix, either peat or a peat substitute. Cover with perlite or vermiculite and place in a cold frame. Germination takes 3–4 weeks. Flowers in the first year.

Lychnis *Caryophyllaceae*

A genus of annuals, biennials, and perennials. Plant in a well-drained soil in a sunny position.

Lychnis alpina Zones 4–8
Tiny seeds: 28,000 per ounce
Hardy perennial
Height 2–6in.
Pale to deep pink flowers with frilled petals
Flowers in summer
Dark green, linear leaves

Sow in spring or autumn in pots or cell packs using standard soil-less seed mix, either peat or a peat substitute. As these are very fine seeds, mix them with the finest sand or talcum powder for an even sowing. Do not cover. Water from the bottom or with a fine spray, place in a cold frame. Germination takes 2–3 weeks. If sowing in autumn, overwinter young plants in the cold frame.

Moltkia *Boraginaceae*

A genus of perennials, evergreens, and deciduous shrubs. Plant in a well-drained, neutral to acid soil, in a sunny situation.

Moltkia petraea Zones 6–9
Medium seeds: 1,540 per ounce
Hardy semievergreen
Height 12 in.
Clusters of violet blue, funnel-shaped flowers
Flowers in summer
Narrow, hairy, long, midgreen leaves

Sow in autumn in pots or cell packs using a standard loam-based seed mix, mixed with coarse horticultural sand (2 parts soil mix + 1 part sand). Cover lightly with coarse sand and place in a cold frame. Germination takes 1–2 months. Overwinter young plants in a cold frame.

Nepeta *Lamiaceae*

A genus of hardy, summer-flowering perennials. Plant in a well-drained soil in a sunny position.

Nepeta racemosa (Catmint) Zones 3–8
Small seeds: 3,640 per ounce
Hardy perennial
Height 18 in.
Small, tubular, lavender-blue flowers in loose spikes
Flowers all summer
Aromatic, grayish-green, small oval leaves with slightly serrated edges

Sow in spring in pots or cell packs using standard, soil-less seed mix, either peat or a peat substitute. Cover with perlite or vermiculite and place under protection at 65°F. Germination takes 14–21 days. Flowers in its first year.

***Recipe for sowing, planting, and growing* Nepeta x faassenii**
The common name of *Nepeta* x *faassenii* is catmint. This is because cats are passionate about it, especially when it is young or a piece of it has been broken. To prevent your cat demolishing the plant and to give the plant a chance to get established, it is a good idea to place an upturned metal/wire mesh hanging basket over the crown. The plant can then grow through the mesh and give you a wonderful display. You will even have enough to give your cat some as a reward.

Ingredients
6 seeds per cell or 10 seeds per pot (approximately)
1 3 x 5 card folded in half
1 flat with cell packs
OR
1 x 4 in. diameter pot
Standard soil-less seed mix, either peat or a peat substitute
Fine-grade perlite (wetted) or vermiculite
White plastic plant label

Method Fill the cell packs or pot with soil-less, smooth over, tap down and water in well. Put a small amount of seed into the crease of the folded card and tap gently in order to sow thinly on the surface of the compost. Press the seed gently into the compost with the palm of the hand. Cover with fine-grade perlite (wetted) or vermiculite. Label with the plant name and date. Place the cell packs or pot in a warm, light place out of direct sunlight at an optimum temperature of 65°F. Keep watering to a minimum until germination has taken place, after 14–21 days in spring with warmth. Shortly after germination, when the seedlings have emerged fully, put the containers into a cooler environment at 59°F. Pot up or prick out approximately four weeks later. If you are using cell packs, you can plant directly into containers. Wait to plant out in the garden until the young plants have had a period of hardening off and there is no threat of frost.

Oenothera *Onagraceae*

A genus of annuals, biennials, and perennials grown for their short-lived, wonderful flowers. Plant in a well-drained soil (can be sandy) in full sun.

Oenothera macrocarpa (missouriensis) **(Missouri evening primrose)**
Zones 4–10
Medium seeds: 560 per ounce
Hardy perennial
Height 4 in.
Lovely, bell-shaped yellow flowers
Flowers throughout the summer
Lance-shaped, midgreen leaves

Sow in spring in pots or cell packs using standard soil-less seed mix, either peat or a peat substitute. Cover with perlite or vermiculite and place in a cold frame or under protection at 70°F. Germination takes three weeks cold, or 8–15 days warm. Flowers in its second season.

Ononis *Papilionaceae*

A genus of summer-flowering perennials and evergreens, and deciduous shrubs grown for their pea-like flowers. Plant in a well-drained soil in a sunny position.

Ononis spinosa **Zones 5–9**
Medium seeds: 476 per ounce
Hardy perennial
Height 10 in.
Rose/purple pea-like flowers
Flowers from summer until early autumn
Small, midgreen, round or oval, three-part leaves

Sow in spring in pots or cell packs using a standard loam-based seed mix. Cover with perlite or vermiculite, place in a cold frame. If using seed more than a year old, you should soak the seed for 12 hours, removing any floating seeds prior to sowing. Germination takes 1–3 months. Flowers during its second season.

Petrorhagia *Caryophyllaceae*

A genus of annuals and perennials suited to a well-drained, sandy soil and a sunny position. These plants self-seed freely.

Petrorhagia saxifraga AGM **(Tunic flower) Zones 4-8**
Small seeds: 4,200 per ounce
Hardy perennial
Height 4 in.
A profusion of cup-shaped, double, white or pale pink flowers, occasionally with deeper pink veins
Flowers in summer
Narrow, grass-like leaves
Looks lovely growing in walls or in a dry rock garden

Sow seeds in autumn in pots or cell packs using a standard, loam-based seed mix. Cover lightly with coarse sand and place in a cold frame. Germination takes 2–4 months. Overwinter young plants in the cold frame. Flowers during its second season.

Phuopsis *Rubiaceae*

A single-species genus, hardy, summer flowering perennial. Plant in a well-drained soil and a sunny situation.

Phuopsis stylosa (Crucianella stylosa) **Zones 6-10**
Medium seeds: 1,400 per ounce
Hardy perennial
Height 12 in.
Sweetly scented clusters of small, tubular, pink/mauve flowers
Flowers all summer
Small, pale green, linear leaves

Sow fresh seeds in autumn in pots or cell packs using a standard, loam-based seed mix. Cover lightly with coarse sand and place in a cold frame. Germination takes 2–4 months. Overwinter young plants in the cold frame. Flowers during its second season.

Platycodon *Campanulaceae*

A genus of hardy perennials grown for their balloon-shaped flower buds. Plant in full sun.

Platycodon grandiflorus AGM **(Balloon flower) Zones 3–8**
Medium seeds: 2,380 per ounce
Hardy perennial
Height up to 2 ft.
Clusters of large, balloon-like buds opening to bell-shaped, blue/purple flowers
Flowers in summer
Oval, lance-shaped, blue/green leaves

Sow fresh seeds in autumn in pots or cell packs using a standard loam-based seed mix. Cover lightly with coarse sand, and place in a cold frame. Germination takes 2–4 months. Overwinter young plants in the cold frame. Flowers during its second season.

Primula *Primulaceae*

A genus of annuals, biennials, perennials, and some evergreens. All have primrose-shaped flowers. There are primulas suitable for almost every type of site—wet, dry, sunny, and shady. Some species dislike wet winters and others dislike full summer heat.

Primula auricula **(Alpine auricula) Zones 3–8**
Small seeds: 8,400 per ounce
Hardy perennial
Height 6–10 in.
Umbels of fragrant, flat yellow flowers
Flowers in spring
Oval, soft, pale green or gray green leaves, which are densely covered with white farina

Sow in spring in pots or cell packs using standard soil-less seed mix, either peat or a peat substitute. Cover with perlite or vermiculite, place in a cold frame, or under protection at 60°F. Germination takes 3–4 weeks cold, or 14–21 days with warmth. Plant in a well-drained soil in full sun or partial shade.

Primula vulgaris **(Primrose) Zones 5–8**
Medium seeds: 2,240 per ounce
Hardy perennial

Height up to 8 in.
Wonderful, scented, open, pale yellow flowers, with a darker eye
Flowers in early spring
Oval, lance-shaped, bright green leaves

Sow fresh seed in late summer in pots or cell packs using standard, soil-less seed mix, either peat or a peat substitute. Cover with perlite or vermiculite, place in a cold frame. Germination takes 2–3 weeks. Can flower in its first season, however there is a better show in the second. Primroses prefer soil that does not dry out and a site in partial shade. It is essential to use fresh seed for easier germination.

Pulsatilla *Ranunculaceae*

A genus of hardy perennials and some evergreens grown for their feathery leaves and bell- or cup-shaped flowers. Plant in a humus-rich, well-drained soil in a sunny situation.

Pulsatilla alpina subsp. *apiifolia* AGM **Zones 3–8**
Medium seeds: 560 per ounce
Hardy perennial
Height up to 12 in.
Bell-shaped, pale yellow flowers
Flowers in spring
Soft, hairy, feathery green leaves

Pulsatilla vulgaris AGM (Pasque flower) **Zones 4–8**
Medium seeds: 1,120 per ounce
Hardy perennial
Height 6–12 in.
Nodding, cup-shaped, flowers in white, blue, pink, or purple
Flowers in spring and early summer
Soft, hairy, feathery leaves

Sow fresh seeds in autumn in pots or cell packs using a standard, loam-based seed mix, mixed with coarse horticultural sand. Mix to a ratio of 2 parts soil mix + 1 part sand. Cover lightly with coarse sand and place in a cold frame. Germination takes 14–28 days, however it can be as long as one year so do not give up. Overwinter young plants in the cold frame. Flowers during its second season.

Roscoea *Zingiberaceae*

A genus of hardy tuberous perennials. Plant in a humus-rich soil, which should not dry out in summer, in sun or partial shade.

Roscoea alpina **Zones 5–7**
Medium seeds: 840 per ounce
Hardy perennial
Height 6–8 in.
Dark purple, hooded, orchid-like flowers
Flowers in summer
Lance-shaped, erect leaves

Sow fresh seed in autumn in pots or cell packs using a standard, loam-based seed mix, mixed with coarse horticultural sand. Mix to a ratio of 1 part soil mix + 1 part sand. Cover lightly with mix and place outside to get all weathers. (see "Breaking Seed Dormancy", page

233, for more information). Germination takes 1–12 months, so do not give up. Flowers in its second year. This plant originates in the Himalayas, where it grows happily at 6–11,000 ft.

***Recipe for sowing, planting, and growing* Roscoea alpina**
This plant is a member of the ginger family. It has truly orchid-like flowers that are stunning in any rock garden.

Ingredients
5 seeds per cell pack or 8 seeds per pot
1 flat with cell packs
OR
1 x 4 in. pot
Standard loam-based seed mix, mixed with coarse horticultural sand.
Mix to a ratio of 1 part soil mix + 1 part sand
Extra coarse horticultural sand to cover the pot
White plastic plant label

Method In autumn, fill the cell packs or pot with soil mix, smooth over, tap down and water in well. Sow the seeds thinly on the surface of the soil mix. Press gently into the mix with the palm of the hand. Cover the seed with coarse horticultural sand. Label with the plant name and date. Place the flat or pot outside, on a level surface, so that it is exposed to all weathers, including frosts.

Do not worry if you live in a snowy area and the containers get immersed in snow, the melting snow will aid germination. If you do not live in an area that will get a winter frost, it is a good idea to put the seed and a handful of damp sand into a clearly marked plastic bag. Seal the bag and place it in the refrigerator for 3 weeks. Remove and sow as mentioned in the first paragraph, then place outside.

Whichever method you use, germination is a bit erratic, taking anything from 1–12 months. Do not give up and discard your soil mix, it might germinate the next week! Whichever method you follow, prick out when the seedlings are large enough to handle. If you are using cell packs, you can plant directly into the garden as soon as the soil is warm enough to dig over prior to planting out.

Sagina *Caryophyllaceae*

A genus of annuals, perennials, and evergreens. Plant in gritty, moist soil in a sunny situation.

Sagina subulata (Heath pearlwort) **Zones 4–8**
Minute seeds: 112,000 per ounce
Hardy perennial
Height 1 1/2 in.
A myriad of tiny, star-shaped, white flowers
Flowers in summer
Bright green mat of foliage

Sow in late summer in pots or cell packs using standard soil-less seed mix, either peat or a peat substitute. As these are very fine seeds, mix them with the finest sand or talcum powder for an even sowing. Do not cover. Water from the bottom or with a fine spray and place in a cold frame. Germination takes 2–3 weeks. Flowers in the following season. Good for planting in between paving stones or as an edging to paths.

Saponaria *Caryophyllaceae*

A genus of annuals and perennials, ideal for rock gardens, screes, and banks. Plant in a well-drained soil and a sunny position.

Saponaria ocymoides (Bouncing Bet) Zones 2–8
Medium seeds: 1,400 per ounce
Hardy perennial
Height up to 3 in.
Creeping habit
Masses of flat, tiny, pale pink to crimson flowers
Flowers all summer
Compact, oval leaves which form a sprawling, hairy mat

Sow in spring in pots or cell packs using standard, soil-less seed mix either peat or a peat substitute. Cover with perlite or vermiculite and place in a cold frame. Germination takes 3–4 weeks. Flowers in the following season.

Saxifraga *Saxifragaceae*

A large genus of hardy to half-hardy perennials and evergreens. Excellent for rock gardens. There are saxifrages suitable for many conditions, from moist to dry soil and from shade to full sun. Check which species you have, so that you can plant it in the ideal situation.

Saxifraga umbrosa (London pride, St. Patrick's cabbage)
Zones 6–9
Tiny seeds: 50,400 per ounce
Hardy perennial
Height 18 in.
Loose panicles of pink starry flowers
Flowers all summer
Rosettes of spatula-shaped, leathery leaves

Sow fresh seed in autumn in pots or cell packs, using a standard, loam-based seed mix, mixed with coarse horticultural sand. Mix to a ratio of 1 part soil mix + 1 part sand. As these are very fine seeds, mix them with the finest sand or talcum powder for an even sowing. Cover lightly with soil mix and place outside exposed to all the weathers. (See "Breaking Seed Dormancy", page 233, for more information). Germination takes 1–12 months, do not give up. Flowers in its second year.

Recipe for sowing, planting, and growing Saxifraga umbrosa

This plant is a very useful for ground cover. It likes being planted in soil that is moist and does not dry out in high summer, but does not get water logged in winter. It prefers partial shade.

Ingredients

20 seeds per cell pack or 30 seeds per pot (or as near as you can manage)
Talcum powder, or fine plain white flour
1 5 x 3 card, folded in half
1 flat with cell packs

OR

1 x 4 in. pot
Standard, loam-based seed mix mixed with coarse horticultural sand.
Mix to a ratio of 1 part soil mix + 1 part sand
Extra coarse horticultural sand to cover the pot
White plastic plant label

Method In autumn, fill the cell packs or pot with compost, smooth over, tap down and water in well. As the seeds are very fine, it is a good idea to mix them with talcum powder or extra fine white flour, in which the seeds will show up. Put a very small amount of the seed mix into the crease of folded 3 x 5 in. card. Tap the card gently to sow the seed thinly. Cover the seed with coarse horticultural sand. Label with the plant name and date. Place the cell packs or pot outside on a level surface, so that it is exposed to all weathers, including frosts. Do not worry if you live in a cold area and the containers get immersed in snow, as the melting snow will aid germination.

If you do not live in an area that will get a winter frost, it is a good idea to put the seed and flour mix, and a handful of damp sand into a clearly marked plastic bag. Seal the bag and place in the refrigerator for 3 weeks. Remove and sow as mentioned in the first paragraph, then place outside.

Whichever method you use, germination is a bit erratic, taking anything from 1 to 12 months. Do not give up and discard your soil mix, it might germinate the next week!

Prick out the seedlings when they are large enough to handle. If you are using cell packs, you can plant the seedlings directly in the garden, as soon as the soil is warm enough to dig over, before planting out.

Scutellaria *Lamiaceae*

A genus of hardy to frost-tender perennials. Plant it in a well-drained soil in a sunny situation.

Scutellaria alpina (Alpine skullcap) Zones 5–8
Medium seeds: 2,520 per ounce
Hardy perennial
Height up to 12 in.
Compact clusters of slender, tubular, hooded flowers, bright blue/violet, shading to almost white just above the calyx
Flowers in summer
Small, mat-forming, downy leaves

Sow in spring in pots or cell packs using standard, soil-less seed mix, either peat or a peat substitute. Cover with perlite or vermiculite and place in a cold frame. Germination takes 4–8 weeks. Occasionally flowers in its first season.

Sedum *Crassulaceae*

A genus of annuals, biennials, and evergreens. Plant in any soil but prefers a fertile, well-drained soil. In sun or partial shade.

Sedum acre (Stonecrop) Zones 3–8
Small seeds: 22,400 per ounce
Hardy evergreen perennial
Height up to 4 in.
Creeping habit
Star-shaped, bright yellow flowers
Flowers in summer
Mat-forming, succulent, bright green leaves

Sedum rupestre L. (*reflexum* L.) (Rock stonecrop, Recurved yellow stonecrop) Zonez 4–9
Tiny seeds: 42,000 per ounce
Hardy, evergreen perennial
Height up to 8 in.
Creeping habit
Clusters of yellow, star-shaped flowers
Flowers in summer
Narrow, fleshy, gray green leaves

Sow fresh seed in autumn in pots or cell packs using a standard, loam-based seed mix, mixed with coarse horticultural sand. Mix to a ratio of 2 parts soil mix + 1 part sand. As these are very fine seeds, mix them with the finest sand or talcum powder for an even sowing. Do not cover. Water from the bottom or with a fine spray and place under protection at 60°F. Germination takes 1–4 months. Overwinter young plants in a frost-free place. Might take up to 3 years to flower.

Silene *Caryophyllaceae*

A genus of annuals, perennials, and some evergreens. Plant in a well-drained soil in a sunny position.

Silene pendula "Compacta" (Nodding catchfly)
Small seeds: 3,360 per ounce
Hardy annual
Height 8 in.
Clusters of double flowers of pink, red, and white
Flowers from summer until early autumn
Midgreen, oval hairy leaves

Sow in spring in pots or cell packs using standard, soil-less seed mix, either peat or a peat substitute. Cover with perlite or vermiculite and place in a cold frame. Germination takes 1–2 weeks.

Sisyrinchium *Iridaceae*

A genus of annuals, perennials, and some semievergreens. This plant will adapt to most conditions, apart from full shade and very wet soils.

Sisyrinchium idahoense var. *bellum* Zones 6–9
Medium seeds: 1,680 per ounce
Hardy, semievergreen perennial
Height up to 5 in.
Pretty, purple/mauve, star-shaped flowers with yellow eyes
Flowers all summer
Bluish-green, tuft- or grass-like foliage

Sow fresh seed in autumn in pots or cell packs using a standard, loam-based seed mix, mixed with coarse horticultural sand. Mix to a ratio of 1 part soil mix + 1 part sand. Cover lightly with soil mix and place in a cold frame. Germination takes 2–4 weeks. Overwinter young plants in a cold frame. Flowers in its second year. Ideal for rock gardens and self-seeds readily.

Symphyandra *Campanulaceae*

A genus of short-lived, hardy perennials. Plant in a well-drained soil and a sunny situation.

Symphyandra hofmannii (Bell flower) Zones 5–9
Small seeds: 3,920 per ounce
Short lived, hardy perennial, sometimes grown as a biennial
Height 2 ft.
Dense spikes of creamy white, nodding, bell-shaped flowers
Flowers in summer
Slightly soft, hairy plant with dense, spiky foliage

Sow in spring in pots or cell packs using standard, soil-less seed mix, either peat or a peat substitute. Cover with perlite or vermiculite and place under protection 68°F. Germination takes 1–2 weeks.

Tellima *Saxifragaceae*

A genus of one species, hardy, semievergreen plants that prefer a well-drained soil and a sunny position. Will also thrive in partial shade beneath shrubs.

Tellima grandiflora (Fringe cups) Zones 4–7
Small seeds: 4,480 per ounce
Hardy, semievergreen perennial
Height 2 ft.
Racemes of lime-green flowers that turn copperish with age
Flowers in summer
Heart-shaped, slightly hairy leaves with a purple tint

Sow fresh seed in autumn in pots or cell packs using a standard, loam-based seed mix, mixed with coarse horticultural sand. Mix to a ratio of 1 part soil mix + 1 part sand. Cover lightly with soil mix and place outside, exposed to all weathers (see "Breaking Seed Dormancy", page 233, for more information). Germination takes 1–12 months; do not give up. Flowers in its second year.

Thymus *Lamiaceae*

A genus of hardy evergreen perennials, all of which have aromatic leaves. Plant in a well-drained soil in a sunny position.

Thymus serpyllum (Mother of Thyme) Zones 4–9
Small seeds: 16,800 per ounce
Hardy, evergreen perennial
Height 1/2 in.
Small, clustered, two-lipped flowers in purple, mauve, or white
Flowers in summer
Very small, oval, aromatic, green leaves

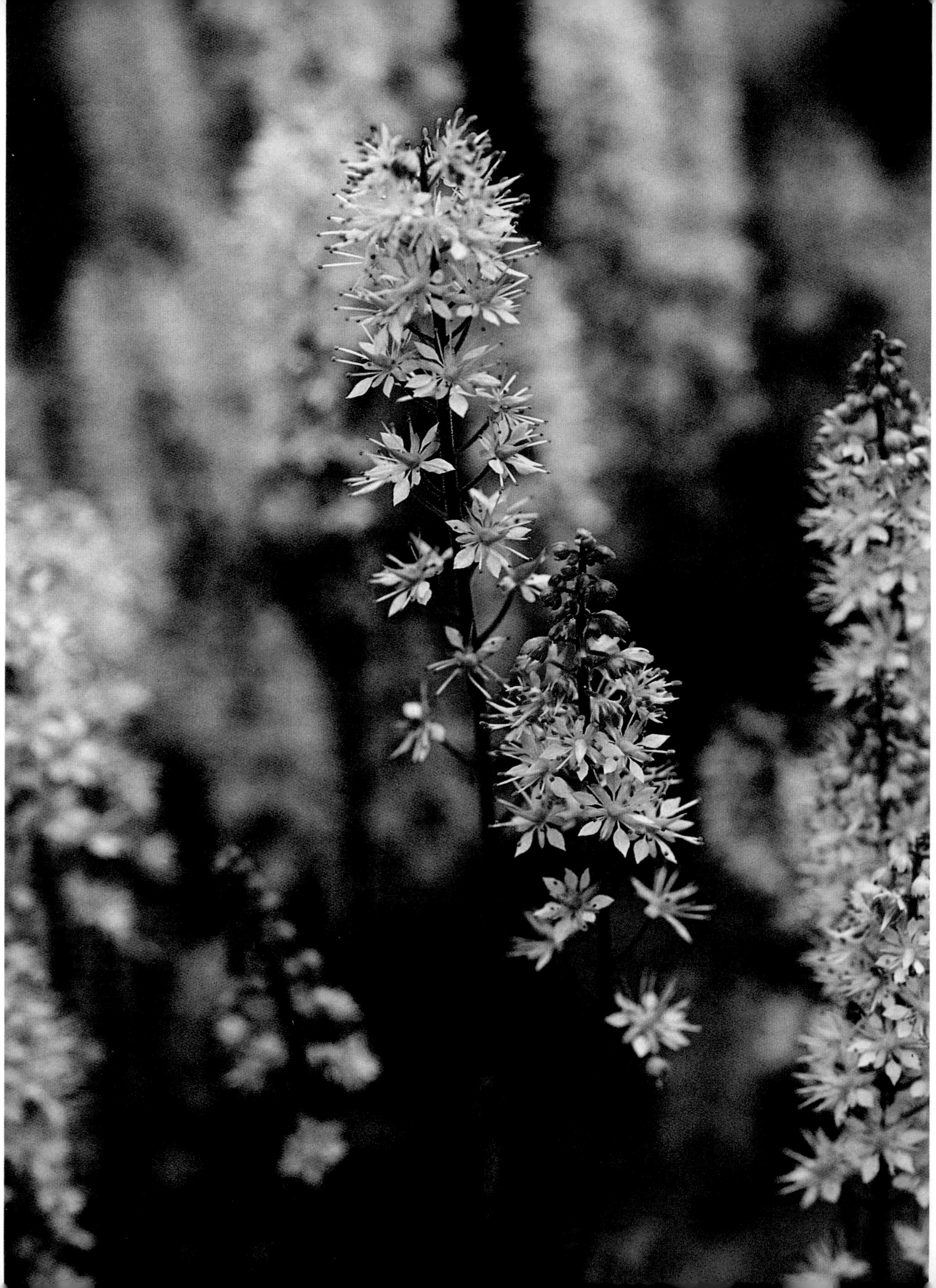

Tiarella wherryi

Sow in spring in pots or cell packs using standard, soil-less seed mix, either peat or a peat substitute. Cover with perlite or vermiculite and place under protection 68°F. Germination takes 1–2 weeks. Do not over water after germination has taken place, as this causes damping off.

Tiarella *Saxifragaceae*

A genus of hardy perennials some of which are evergreen. Plant in a well-drained soil in partial shade.

Tiarella wherryi AGM (Foam flower) Zones 3–8
Small seeds: 5,040 per ounce
Hardy perennial
Height 4 in.
Racemes of tiny, star-shaped, soft pink or white flowers
Flowers from late spring to early summer
Small, hairy, midgreen leaves

Sow fresh seed in autumn in pots or cell packs using a standard, loam-based seed mix, mixed with coarse horticultural sand. Mix to a ratio of 1 part soil mix + 1 part sand. As these are very fine seeds, mix them with the finest sand or talcum powder for an even sowing. Cover lightly with mix, place outside exposed to all weathers (see "Breaking Seed Dormancy", page 233, for more information). Germination takes 1–12 months, so do not give up. Flowers in its second year.

Veronica *Scrophulariaceae*

A genus of hardy perennials, some of which are evergreen, grown for their blue flowers. Plant in a well-drained soil in a sunny position.

Veronica repens (Creeping speedwell) Zones 4–8
Small seeds: 19,600 per ounce
Hardy perennial
Height 2 in.
Creeping habit
Numerous, bluish-white flowers
Flowers in late summer
Vivid green, mat-forming leaves

Sow in spring in pots or cell packs using standard, soil-less seed mix, either peat or a peat substitute. As these are very fine seeds, mix them with the finest sand or talcum powder for an even sowing. Do not cover. Water from the bottom or with a fine spray, place under protection at 68°F. Germination takes 1–2 weeks.

Viola *Violaceae*

A genus of annuals, perennials, and some semievergreens, all grown for their attractive flowers. Plant in a well-drained, but moisture retentive, soil in sun or partial shade.

Viola "Queen Charlotte" Zones 6–9
Medium seeds: 2,800 per ounce
Hardy perennial
Height 6 in.

Dark blue, traditional violet-shaped flowers, which are richly scented
Flowering from early spring
Heart-shaped, midgreen leaves

Sow fresh seed in autumn in pots or cell packs using a standard, loam-based seed mix, mixed with coarse horticultural sand. Mix to a ratio of 1 part soil mix + 1 part sand. As the shells of these seeds are hard, it is a good idea to scarify with fine sandpaper before sowing (see page 233). Then place outside, exposed to all weathers (see "Breaking Seed Dormancy", page 233, for more information). Germination takes 1–4 months. Flowers in its second year.

***Recipe for sowing, planting, and growing** Viola "Queen Charlotte"*
The wonderful scent of the sweet violet makes it worth growing in the garden, and the attractive spring flowers are another bonus. If you do not have space in the garden, it looks lovely in a container, which can be left outside for the whole season.

Ingredients
6 seeds per cell or 10 seeds per pot (or as near as you can manage)
1 sheet of fine sandpaper, cut in half
1 piece white 3 x 5 card, folded in half
1 flat with cell packs
OR
1 x 4 in. pot
Standard loam-based seed mix, mixed with coarse horticultural sand. Mix to a ratio of 1 part soil mix + 1 part sand
White plastic plant label

Method In autumn, fill the cell packs or pot with soil mix, smooth over, tap down and water in well. Select a small amount of seed, place it on a half sheet of fine sandpaper and cover it with the other half. Hold between both hands and slide the sheets of sandpaper back and forth, gently scratching the surface of the seed. Alternatively, you can simply put the other half of the sandpaper on top of the seed and gently push the paper up and down, this will have the same effect. Put a small amount of this scarified seed into the crease of the 3 x 5 folded card. This will allow you to sow the seed thinly on the surface of the compost. Cover with coarse horticultural sand. Label with the plant name and date. Place the flat or pot outside, on a level surface, so it is exposed to all weathers, including frosts. Do not worry if you live in a cold area and the containers get immersed in snow. The melting snow will aid germination.

If you do not live in an area which will get a winter frost, it is a good idea to put the scarified seed and a handful of damp sand into a plastic bag, which is clearly marked. Seal the bag and place in a refrigerator for three weeks. Remove and sow as described above. Then place outside. Germination will occur in the following spring, and flowering the year after that.

Prick out when the seedlings are large enough to handle. If you are using cell packs, you can plant directly in the garden as soon as the soil is warm enough to dig over.

annuals & biennials

It is natural for annuals to reproduce from seed, making your task easier. The crucial thing is to care for the seedlings properly once they have germinated. Planting seedlings out too late or in unsuitable growing conditions will trigger a panic reaction. Programed to survive at all costs, the plant will go into flower rapidly, producing seeds to ensure its survival the following year. Bear in mind as well what will happen to the plant when it dies back. If it is a self-seeder, ask yourself whether allowing the plant to seed freely will cause problems the following year. Once you have grasped these points, you will be free to enjoy your annuals, revelling in their magnificent displays of color and texture and their glorious scents.

Ageratum *Asteraceae*

A genus of half-hardy annuals, and biennials, which attract bees and butterflies. Plant in fertile, well-drained soil in a sunny situation. Do not allow the plants to dry out, as this will inhibit flowering.

Ageratum houstonianum "Blue Danube" (**Floss Flower**)
Small seeds: 18,760 per ounce
Half-hardy annual
Height 6 in.
Clusters of feathery, midblue flowers
Flowers from early summer to frost
Oval, midgreen leaves

As these are very fine seeds, mix them with the finest sand or talcum powder for an even sowing. Sow in spring in pots or cell packs using a soil-less seed mix, either peat or peat substitute. Do not cover. Water from the bottom of the pot or use a fine spray. Place under protection at 60°F. Germination takes 8–10 days.
OR
Sow in prepared, open ground in late spring when air temperature does not go below 48°F at night. Germination will take 14–20 days.

Alcea *Malvaceae*

A genus of hardy biennials (sometimes grown as annuals), and short-lived perennials. All have attractive spikes of flowers. Plant in free-draining soil in a sunny situation.

Alcea rosea Chater's Double Group (**Hollyhock**) Zones 3–9
Medium seeds: 126 per ounce
Hardy biennial
Height 6 ft.
Attractive spikes of rosette-like double flowers in varying colors from white to yellow, pink, and maroon
Flowers from summer to early autumn
Lobed, midgreen leaves

Alcea rosea "Nigra" Zones 3–9
Medium seeds: 126 per ounce
Hardy biennial
Height 5 ft.
Attractive spikes of rosette-like, double flowers in varying colors from deep maroon to near black
Flowers from summer to early autumn
Lobed, mid-green leaves

Sow in spring or late summer in pots or cell packs using a soil-less seed mix, either peat or peat substitute. Cover with perlite or vermiculite and place under protection at 65°F. Germination takes 6–10 days.

Amaranthus *Amaranthaceae*

A genus of half-hardy annuals grown for their colorful foliage and dense clusters of tiny flowers. Plant in well-drained soil in a sunny situation.

Amaranthus caudatus (**Love-lies-bleeding**)
Small seeds: 4,200 per ounce
Half-hardy annual
Height 4 ft.
Red flowers clustered in attractive, hanging tassels
Flowers from summer to early autumn
Pale green, oval leaves with pointed tips

Sow in spring in pots or cell packs using a soil-less seed mix, either peat or peat substitute. Cover with perlite or vermiculite and place under protection at 60°F. Germination takes 8-14 days.

Ammi *Apiaceae*

A genus of hardy annuals and perennials, which have attractive white flowers. These plants will adapt to most gardens, but they prefer a well-drained soil and a sunny situation.

Ammi majus (**Bishop's flower, Queen Anne's Lace**)
Small seeds: 3,640 per ounce
Hardy annual
Height 3 1/2 ft.
Large clusters of lacy white flowers
Flowers all summer
Small oblong mid green leaves

Sow fresh seed in late spring in prepared, open ground when the air temperature does not fall below 50°F at night. Germination takes 14–21 days.

Ammobium *Asteraceae*

A genus of Australian annuals with "everlasting" flowers. Plant in a well-drained, sunny situation.

Ammobium alatum
Medium seeds: 2,800 per ounce
Half-hardy annual
Height 3 ft.
Silvery-white, "everlasting" flowers with yellow centers
Flowers in the summer
Oblong, midgreen leaves

Sow in spring in pots or cell packs using a soil-less seed mix, either peat or peat substitute. Cover with perlite or vermiculite and place under protection at 65°F. Germination takes 5–7 days.

Anagallis *Primulaceae*

A genus of hardy annuals and creeping perennials. Plant in a fertile, moist soil in an open, sunny situation.

Anagallis arvensis var. *caerulea* (**Pimpernel, Scarlet pimpernel**)
Small seeds: 3,920 per ounce
Hardy annual
Height 8 in.

The lacy white flowers of *Ammi majus* with the pink flowers of *Lavatera* "Bredon Springs"

Blue flowers that close in the evening
Flowers all summer
Small, bright green, oval leaves

Sow in spring in pots or cell packs using a soil-less seed mix, either peat or peat substitute. Cover with perlite or vermiculite and place under protection at 65°F. Germination takes 10–14 days.

Antirrhinum *Scrophulariaceae*

A genus of hardy annuals, perennials, and semievergreen subshrubs which flower from spring until autumn. Regular dead-heading will prolong flowering. Plant in a rich, well-drained soil in a sunny situation.

Antirrhinum majus (Snapdragon)
Small seeds: 15,400 per ounce
Perennial grown as half-hardy annual
Height 8–18 in.
Attractive spikes of 2-lipped, sometimes double flowers in white, pink, red, purple, or yellow
Flowers from summer until hard frost
Lance-shaped, midgreen leaves

Sow in spring in pots or cell packs using a soil-less seed mix, either peat or peat substitute. Cover with perlite or vermiculite and place under protection at 65°F. Germination takes 7–14 days.
OR
Sow in late spring in prepared, open ground when the air temperature does not drop below 40°F at night. Germination takes 14–21 days.

Arctotis *Asteraceae*

A genus of tender annuals and perennials. Grow at a minimum temperature of 35–40°F. Plant in a well-drained soil and full sun.

Arctotis "Harlequin"
Medium seeds: 364 per ounce
Half-hardy annual
Height 20 in.
Daisy-like, multi-colored flowers, ranging from red to cream with lovely blue centers
Flowers from summer until early autumn
Chrysanthemum-shaped, dark green leaves with gray undersides

Arctotis stoechadifolia (Blue-eyed African daisy)
Medium seeds: 420 per ounce
Half-hardy perennial often grown as an annual
Height 20 in.
Daisy-like, creamy-white flowers with lovely blue centers
Flowers from summer until hard frost
Chrysanthemum-shaped, dark green leaves with gray undersides

Sow in spring in pots or cell packs using a soil-less seed mix, either peat or peat substitute. Cover with perlite or vermiculite and place under protection at 70°F. Germination takes 4–8 days.

Bassia *Chenopodiaceae*

A genus of annuals and perennials with striking foliage. Plant in fertile, well-drained soil in a sunny situation. Stake on windy sites.

Bassia scoparia f. *trichophylla* AGM (*Kochia scoparia* f. *trichophylla*) (Burning bush)
Medium seeds: 2,800 per ounce
Half-hardy annual
Height 3 ft.
Insignificant, small flowers
Flowers in summer
Light green foliage that turns deep red in autumn

Sow in spring in pots or cell packs using a soil-less seed mix, either peat or peat substitute. Cover with perlite or vermiculite and place under protection at 68°F. Germination takes 4–10 days.

Begonia *Begoniaceae*

A large genus which includes annuals, perennials, evergreens, deciduous shrubs, and small trees. Grown for their attractive flowers or foliage. Plant in a slightly acid soil in semishade to full sun.

Begonia semperflorens Cultorum Group (Wax begonia)
Seeds so small they appear as dust: 196,000 per ounce
Half-hardy evergreen
Height 10–12 in.
Small, pink flowers, sometimes with a white center
Flowers from spring until frost
Dark green, round, fleshy leaves with a light green underside

There are many named hybrids available from seed companies, all with different attributes. For example, *Begonia semperflorens* "Cocktail" has pink, red or white flowers with waxy bronze leaves.

Begonia sutherlandii
Seeds so small they appear as dust: 168,000 per ounce
Height trailing to 3 1/2 ft.
Clusters of small orange flowers
Flowers in summer
Lance-shaped, bright green leaves with red veins
Good for hanging baskets or cascading over tubs

Sow in spring in pots or cell packs using a soil-less seed mix, either peat or peat substitute. As these are very fine seeds, mix them with the finest sand or talcum powder for an even sowing. Do not cover. Water from the bottom or with a fine spray, place under protection at 70°F. Germination takes 14–28 days.

Bellis *Asteraceae*

A genus of hardy perennials, some cultivars are grown as biennials, all are grown for their daisy-like flowers. Plant in fertile, well draining soil in sun to partial shade.

Bellis perennis (English Daisy) Zones 3–9
Small seeds: 15,400 per ounce
Hardy perennial grown as a biennial
Height 16 in.
Attractive white or pink double flowers
Flowers from late spring to early summer
Oval, midgreen leaves

Sow in summer in cell packs using a soil-less seed mix, either peat or peat substitute. Cover with perlite or vermiculite and place in a cold frame. Germination takes 2–3 weeks. Overwinter the young plants in the cold frame. They will flower in their second year.
OR
Sow in spring or autumn in pots or cell packs using a soil-less seed mix, either peat or peat substitute. Cover with perlite or vermiculite, place under protection at 70°F. Germination takes 7–14 days in spring and 14–28 days in autumn. The plants grown from spring-sown seeds will flower the following year and those from autumn-sown seeds will flower the following spring.

Recipe for sowing, planting, and growing Bellis perennis

English daisies are cheerful plants to have in the garden. Cultivated varieties look great in containers or at the front of a border. It is worth dead-heading after the first flush of flowers to prolong the flowering season of the plants.

Bellis perennis seeds can be sown at three different times of year. If you start your seeds off in spring, you will grow a hardy plant that can be planted out and overwintered in the garden, to flower the following spring. Seeds sown in the summer can be germinated and overwintered in the cold frame. Alternatively, you can sow in the autumn. Plants will have to be overwintered in a frost-free place. Watch out for rot. There is a bonus to autumn planting—your plants will flower in their first season. Whenever you start your sowing, the method is basically the same.

Ingredients
10 seeds per cell or 15 seeds per pot
1 3 x 5 card, folded in half
1 flat with cell packs
OR
1 x 4 in. pot
Soil-less seed mix, either peat or peat substitute
Fine-grade perlite (wetted) or vermiculite
White plastic plant label

Method Fill the cell packs or pot with soil-less mix, smooth over, tap down and water thoroughly. As the seeds are so fine, it is worth putting a very small quantity of seed into the crease of a folded 3 x 5 card. You can then sow the seed by tapping the card gently. This will let you see what you are sowing and allow you to sow thinly on the surface of the soil-less mix. Cover the seed with perlite or vermiculite and then label clearly with the plant name and the date.

For spring sowings, place the cell packs or pot in a warm, light place out of direct sunlight, at an optimum temperature of 70°F. Summer-sown seed should be placed in a cold frame or a cool greenhouse. Keep the mix damp, but not wet, until the seed germinates. This takes 7–14 days in spring if the seed is put in a warm place. Summer germination in a cold frame takes 14–21 days and seed given some warmth in autumn will germinate after 14–28 days.

Put pots of spring and summer sown seeds in a cooler environment of about 59°F shortly after germination has taken place and the seedlings have emerged fully. You can prick out and pot up approximately 2–3 weeks after germination. The spring seedlings will be ready to plant out approximately 4 weeks after that. If you are using cell packs, you can plant out direct in the garden after a period of hardening off and when there is no threat of frost.

The summer seedlings should be left in pots to overwinter in the cold frame. Autumn-sown plants can be pricked out at 4–8 weeks, depending on light levels after germination, potted up and over-wintered with frost protection. Alternatively, if you have sown thinly and the seedlings are not overcrowded, you can wait until early spring for potting up or planting out. Whichever way you choose, you can plant out in the garden as soon as all threat of frost has passed.

Bidens *Asteraceae*

A genus of hardy annuals and perennials. Plant in a well-drained soil in a sunny position.

Bidens ferulifolia "Goldie" (Spanish needles)
Small seeds: 3,360 per ounce
Hardy annual
Height 18 in. trailing habit
Attractive golden flowers
Flowers throughout the summer
Mid-green, serrated leaves
This variety is good for hanging baskets or tubs

Sow in early spring in pots or cell packs using a soil-less seed mix, either peat or peat substitute. Cover with perlite or vermiculite, place under protection at 68°F. Germination takes 14–21 days.

Brachycome *Asteraceae*

A genus of hardy annuals and perennials with daisy-like flowers. Plant in a fertile, well-drained soil in a sunny, sheltered position.

Brachycome iberdifolia (Swan River daisy)
Small seeds: 14,000 per ounce
Hardy annual
Height 18 in.
Small blue (sometimes pink, mauve, purple, or white) fragrant, daisy-like flowers
Flowers in summer
Deeply cut, midgreen leaves

Sow in early spring in pots or cell packs using a soil-less seed mix, either peat or peat substitute. Cover with perlite or vermiculite, place under protection at 65°F. Germination takes 10–14 days.

Browallia *Solanaceae*

A genus of tender perennials usually grown as annuals in northern climates. Plant in a fertile, well-drained soil in partial shade.

Browallia speciosa "Starlight"
Small seeds: 11,200 per ounce
Half-hardy perennial, grown as an annual
Height 6 in.
Attractive, violet blue flowers with white eyes
Flowers all summer until frost
Oval, midgreen leaves

Sow in early spring in pots or cell packs using a soil-less seed mix, either peat or peat substitute. Cover with perlite or vermiculite and place under protection at 72°F. Germination takes 14–21 days. In controlled conditions this plant will flower from seed in about 90 days. It looks lovely in hanging baskets or patio containers.

Bupleurum *Apiaceae*

A genus of hardy annuals, perennials, and evergreen shrubs. Plant in a well-drained soil in full sun. These plants thrive in coastal gardens.

Bupleurum griffithii (Thoroughwax)
Medium seeds: 840 per ounce
Hardy annual
Height 2 1/2 ft.
Umbels of green/yellow flowers
Flowers in summer
Eucalyptus-shaped leaves
The foliage is lovely as a filler in flower arrangements. May need staking in exposed sites

Sow in late spring in prepared, open ground when air temperature does not go below 50°F at night. This plant is best grown where it is to flower. Germination takes 14–21 days.

Calandrinia *Portulacaceae*

A genus of fleshy, hardy to tender, annual or perennial, spreading or trailing plants.

Calandrinia umbellata
Small seeds: 14,000 per ounce
Tender perennial grown as an annual
Height 6 in.
Bright crimson-magenta, cup-shaped flowers
Flowers in early summer
Linear, hairy, midgreen leaves

Sow in early spring in pots or cell packs using a soil-less seed mix, either peat or peat substitute. Cover with perlite or vermiculite, place under protection at 65°F. Germination takes 7–14 days. Flowers close at night and when the weather is cloudy. Requires cool temperature to grow well.

Calceolaria *Scrophulariaceae*

A genus of hardy to frost-tender annuals, biennials, and evergreen perennials. Plant in well-drained soil, in a sunny site with a bit of protection from the midday sun. Dislikes cold and wet, especially intolerant of wet winters.

Calceolaria integrifolia (*rugosa*) (Slipper flower)
Medium seeds: 2,380 per ounce
Evergreen subshrub often grown as an annual
Height 9 in.
Pouch-shaped, yellow/reddish-brown flowers
Flowers in summer
Elliptic, midgreen leaves

Sow in early spring in pots or cell packs using a soil-less seed mix, either peat or peat substitute. Cover with perlite or vermiculite and place under protection at 68°F. Germination takes 7–10 days. Sowing in late spring will produce flowering plants in early autumn, so protect from early frosts.

Callistephus *Asteraceae*

A one-species genus, half-hardy annual. Plant in a fertile, well-drained soil in a sunny and sheltered site. There are many named forms of this flower, available in a full range of colors.

Callistephus chinensis (China aster)
Medium seeds: 1,148 per ounce
Half-hardy annual
Height 6in–3 ft.
The flowers vary from daisy to chrysanthemum, double to single in shades of white, pink, and blue
Flowers from summer until early autumn
Oval, toothed, midgreen leaves

Callistephus chinensis "Matador mix"
Medium seeds: 1,400 per ounce
Half-hardy annual
Height 3 ft.
Large chrysanthemum-like flower heads, slightly in-curved, in shades of pink, white, and purple
Flowers from summer until early autumn
Oval, midgreen toothed leaves
This variety is good for cut flowers

Sow in spring in pots or cell packs using a soil-less seed mix, either peat or peat substitute. Cover with perlite or vermiculite, place under protection at 70°F. Germination takes 8–10 days.
OR
Sow in late spring in prepared, open ground when air temperature is at least 50°F at night. Germination takes 14–21 days.

Carthamus *Asteraceae*

A genus of hardy annuals. Plant in well-drained soil in a sunny situation.

Carthamus tinctorius (Safflower)
Medium seeds: 56 per ounce
Annual
Height 3 ft.
Lovely, yellow or orange, thistle-like flowers in summer
Oval, spine-toothed leaves

Dries well for flower arrangements
Also used as a dye plant

Sow in late spring in prepared, open ground, when air temperature does not go below 50°F at night. Germination takes 14–21 days.

Celosia *Amaranthaceae*

A genus of half-hardy perennials often grown as annuals. Plant in fertile, well-drained soil in a sunny, sheltered situation.

Celosia argentea var. *cristata* (Cockscomb, Crested celosia)
Medium seeds: 2,800 per ounce
Tender perennial, often grown as an annual
Height 1–2 ft.
Unusual, pyramid-shaped, feathery flower heads in red, yellow, pink or apricot
Oval, midgreen leaves

Sow in spring in pots or cell packs using a soil-less seed mix, either peat or peat substitute. Cover with perlite or vermiculite and place under protection at 65°F. Germination takes 8–10 days.
OR
Sow in late spring in prepared, open ground when the air temperature does not drop below 50°F at night. Germination takes 14–21 days.

Centaurea *Asteraceae*

A genus of hardy annuals and perennials all grown for their attractive flowers. Plant in well-drained soil and a sunny situation.

Centaurea cyanus (Cornflower, Bachelor Button)
Medium seeds: 560 per ounce
Hardy annual
Height 2 ft.
Lovely, daisy-like flowers in shades of blue, pink, white or purple
Lance-shaped, gray-green leaves
Excellent for cutting

Centaurea cyanus

Cerinthe major "Purpurascens"

Sow in spring in prepared, open ground when the air temperature does not go below 50°F at night. Germination takes 14–21 days.

Cerinthe *Boraginaceae*

A genus of hardy annuals and perennials grown for their interesting foliage. Plant in fertile soil that retains a certain amount of moisture. Will grow happily in a clay soil in a sunny situation.

Cerinthe major "Purpurascens" (Blue Honeywort, Blue Shrimp Plant)
Medium seeds: 168 per ounce
Hardy annual
Height 2 ft.
Attractive, deep blue bracts around rich purple blue, bell-shaped flowers tinged with cream
Flowers from late spring and through the summer
Handsome, glaucous, green-blue leaves
Attractive to bees

Sow in spring in pots or cell packs using a soil-less seed mix, either peat or peat substitute. Cover with perlite or vermiculite, place under protection at 65°F. Germination takes 8–10 days.
OR
Sow in late spring in prepared, open ground when air temperature does not go below 50°F at night. Germination takes 14–21 days.

Chenopodium *Chenopodiaceae*

A very interesting genus of annuals, perennials, and subshrubs, widely distributed throughout the world. These plants will adapt to most situations. The species mentioned below should be planted in free-draining soil in a sunny situation.

Chenopodium botrys (Jerusalem Oak, Ambrosia Jerusalem Oak)
Small seeds: 16,800 per ounce
Half-hardy annual
Height 1–3ft
Spikes of feathery, aromatic small flowers
Flowers in summer until early autumn
Hairy, aromatic, pinnate, pale green leaves

Sow in late spring in prepared, open ground, when air temperature is at least 48°F at night. Germination takes 14–21 days. Flowers are long lasting which makes them suitable for dried-flower arrangements.

Clarkia *Onagraceae*

A genus of hardy annuals, which are grown for their flowers. Plant in a fertile, well-drained soil in a sunny situation.

Clarkia amoena (Satin flower, Farewell to spring)
Medium seeds: 2,800 per ounce
Hardy annual
Height 2 ft.
Attractive spikes of single or double flowers in shades of pink
Flowers in summer where summers are cool
Lance-shaped, midgreen leaves
These plants look lovely if grown in large clumps

Clarkia "Brilliant"
Small seeds: 5,600 per ounce
Hardy annual
Height 2ft
Attractive, double rosettes of bright pink or red flowers in long spikes
Flowers in summer
Oval, midgreen leaves

Sow in late spring in prepared, open ground when air temperature does not go below 48°F at night. Germination takes 14–21 days.

Cleome *Capparaceae*

A genus of half-hardy annuals and evergreen shrubs, all grown for their unusual, spidery flowers. Plant in fertile, well-drained soil in a sunny position.

Cleome hassleriana "Cherry Queen" (Spider flower)
Medium seeds: 1,120 per ounce
Half-hardy annual
Height 3 ft.
Large, open, airy, bright rose-colored, narrow petaled flowers with strong scent
Flowers all summer until the first frosts
Mid-green, divided, lance-shaped leaves

Sow in spring in pots or cell packs using a soil-less seed mix, either peat or peat substitute. Cover with perlite or vermiculite and place under protection at 70°F. Germination takes 8–14 days; however, germination can be very erratic and very seasonal.

Coreopsis *Asteraceae*

A genus of annuals and perennials, which have daisy-like flowers. Plant in well-drained soil in a sunny situation.

Coreopsis tinctoria (Caliopsis)
Medium seeds: 840 per ounce
Hardy annual
Height 2–3 ft.
Large, daisy-like, bright yellow flowers with red veins
Flowers throughout the summer until early autumn
Lance-shaped, midgreen leaves

Sow in spring in pots or cell packs using a soil-less seed mix, either peat or peat substitute. Cover with perlite or vermiculite, place under protection at 65°F. Germination takes 8–10 days.
OR
Sow in late spring in prepared, open ground when air temperature is at least 50°F at night. Germination takes 14–21 days.

Cosmos *Asteraceae*

A genus of hardy and half-hardy annuals and perennials, which are grown for their attractive flowers. Plant in moist, but not wet, soil in a sunny position. Never let the plants dry out completely.

Eschscholzia *Papaveraceae*

A genus of hardy annuals grown for their poppy-like flowers. Plant in a well-draining soil in a sunny situation. Deadhead regularly to prolong flowering.

Eschscholzia californica AGM (California poppy)
Medium seeds: 1,960 per ounce
Hardy annual
Height 12 in.
Lovely, cup-shaped flowers in shades of yellow, orange, rose, mahogany, and ivory
Flowers all summer until the first frosts
Feathery, blue-green leaves

Sow in spring in prepared open ground when air temperature does not go below 48°F at night. Germination takes 14–21 days.

Eustoma *Gentianaceae*

A genus of halfhardy annuals and perennials with attractive flowers, a mixture of a rose and a poppy. Plant in a well-drained soil in a sunny situation. In cooler climates they make very good potted plants.

Eustoma grandiflorum (*russellianum*) (syn. *Lisianthus*)
Minute seeds: 2,058 per ounce
Halfhardy annual
Height 18in
Poppy or rose-like flowers in pink, blue or white
Flowers all summer
Lance-shaped, green leaves
Good for cut-flower arrangements

Sow in spring in pots or cell packs using a soil-less seed mix, either peat or peat substitute. Do not cover the seed. Place under protection at 70°F. Germination takes14–21 days.

Exacum *Gentianaceae*

A genus of tender annuals, biennials, and perennials, with a minimum growing temperature of 45°F, grown for their mass of flowers. These plants make lovely potted plant. For the garden, plant in a well-drained soil in a sunny situation.

Exacum affine (Persian violet)
Minute seeds: 98,000 per ounce
Tender, evergreen biennial grown as an annual
Height 8in
Tiny, scented, saucer-shaped purple flowers with a yellow eye
Flowers in summer and early autumn
Oval, glossy leaves

Sow in early spring or late summer in pots or cell packs using a soil-less seed mix, either peat or peat substitute. As these are very fine seeds, mix them with the finest sand or talcum powder for an even sowing. Do not cover. Water from the bottom or with a fine spray and place under protection at 70°F. Germination takes 14–21 days. Spring-sown plants will flower the following summer.

Felicia *Asteraceae*

A genus of hardy to frost-tender annuals, perennials, evergreens, and subshrubs. Grown for their blue, daisy-like flowers. Plant in well-drained soil in a sunny situation. These plants do not like being overwet.

Felicia bergeriana (Kingfisher daisy)
Medium seeds: 2,240 per ounce
Hardy annual
Height 6 in.
Small, blue, daisy-like flowers with yellow centers, which close at night and on cloudy days
Flowers in summer and early autumn
Lance-shaped, hairy, green gray leaves

Sow in early spring in pots or cell packs using a soil-less seed mix, either peat or peat substitute. Cover with perlite or vermiculite and place under protection at 68°F. Germination takes 7–14 days.

Gomphrena *Amaranthaceae*

A genus of halfhardy annuals, biennials, and perennials. Plant in well-drained, fertile soil in a sunny situation.

Gomphrena globosa (Globe amaranth)
Medium seeds: 1,120 per ounce
Halfhardy annual
Height 10 in.
Small, clover-like flower heads in pink, purple, white, red, and orange
Flowers in summer until frost
Good for dried-flower arrangements
Hairy, light green, oblong leaves

Sow in early spring in pots or cell packs using a soil-less seed mix, either peat or peat substitute. Cover with perlite or vermiculite and place under protection at 65°F. Germination takes 12–14 days.

Helianthus *Asteraceae*

A genus of annuals and perennials, which are grown for their large, daisy-like flowers that follow the sun with their flower heads. Plant in well-drained soil in a sunny position. The tall varieties will need staking in exposed, windy sites.

There are many cultivars of sunflower now available—tall, short, multicolored, or even double flowered. The seed count varies according to cultivar, so I have taken an average count and given a general description.

Helianthus annuus (Sunflower)
Medium seeds: 84 per ounce
Hardy annual
Height 1–10 ft.
Lovely, large, yellow flowers with a brown or purplish center
Flowers in summer
Large, oval, serrated, midgreen leaves

Helianthus annuus

Sow in spring in prepared open ground when air temperature does not go below 50°F at night. Germination takes 14–21 days.
OR
Sow dwarf varieties, using 2 seeds per 6 in. pot. Use a soil-less seed mix, either peat or peat substitute. Cover lightly with soil mix, place under protection at 55°F. Germination takes 6–12 days.

> ### *Recipe for sowing, planting, and growing* **Helianthus annuus**
> Sunflowers are as good as their name, it is wonderful to watch their heads turn to follow the sun. There are many stunning varieties now available, short ones and tall ones, in colors ranging from yellow to orange, red, and brown. All of them can be grown from seed, which is easy to handle. Sunflowers are a great way to introduce children to gardening. Even when the flower is over, the seed head is spectacular and a great favorite with wild birds, such as finches and chickadees.
>
> When planting in the garden, choose a sunny site with well-drained soil. If your site is exposed, make sure you give the plants some shelter from the prevailing wind, it is also a good idea to stake the tall plants because the flower heads do become very heavy and are liable to snap the stem. They look stunning when planted around the base of a tree or in large clumps against a dark fence.
>
> #### *Ingredients*
> 2 seeds per pot
> 1 x 6 in. diameter pot
> Soil-less seed mix, either peat or peat substitute
> White plastic plant label
> **OR**
> Prepare a site in the garden
>
> #### *Method* Fill the pot with soil mix, smooth over, tap down and water well. Sow the seeds by placing them in the pot well apart and not too close to the edge. Press gently into the soil mix with the palm of the hand. Cover lightly with soil mix, and label with the plant name and date. Place the pot in a warm, light place, out of direct sunlight at an optimum temperature of 55°F. Keep watering to a minimum until germination has taken place, which takes 6–12 days in spring with warmth. The seedling will be ready to plant out or pot up approximately 2–3 weeks after germination.
> **OR**
> Sow directly in a prepared site in the garden when the night-time temperature does not fall below 50°F. Space the seeds 12–18 in. apart, press gently into the soil, cover lightly, and water in well. Germination takes 14–21 days.

Helichrysum *Asteraceae*
A genus of hardy to frost-tender annuals, perennials, and evergreens. Plant in a well-drained soil in a sunny situation.

Helichrysum bracteatum Monstrosum Series **(Strawflower)**
Small seeds: 4,200 per ounce
Half-hardy annual
Height 3 ft.
Papery, double flowers in white, pink, red, orange, and yellow which dry very well
Flowers throughout summer until frost

Sow in early spring in pots or cell packs using a soil-less seed mix, either peat or peat substitute. Cover with perlite or vermiculite and place under protection at 65°F. Germination takes 14–18 days.
OR
Sow in spring in prepared, open ground when air temperature does not go below 50°F at night. Germination takes 21–28 days.

Heliotropium *Boraginaceae*
A genus of hardy to frost-tender annuals and evergreen plants. Plant in well-drained, fertile soil and a sunny position.

Heliotropium arborescens "Marine" **(Heliotrope)**
Small seeds: 4,480 per ounce
Half-hardy, evergreen perennial often grown as an annual in cool climates
Height 18 in.
Small, sweet-scented, deep violet-blue flowers in dense clusters
Flowers in the summer
Dark green, lance-shaped slightly wrinkled leaves

Sow in early spring in pots or cell packs using a soil-less seed mix, either peat or peat substitute. Cover with perlite or vermiculite and place under protection at 60°F. Germination takes 14–21 days.

Helipterum *Asteraceae*
A genus of half-hardy annuals and perennials. Plant in poor, very well-drained soil in a sunny situation.

Helipterum manglesii (syn. *Rhodanthe manglesii*)
Medium seeds: 1,680 per ounce
Half-hardy annual
Height 12 in.
Papery, daisy-like flowers in a variety of colors, red, pink, or white
Flowers from late spring until the first frosts
Pointed, oval, gray green leaves

Helipterum roseum (syn. *Acroclinium roseum*) **(Strawflower)**
Medium seeds: 840 per ounce
Half-hardy annual
Height 12 in.
Small, pink, papery, daisy-like flowers
Flowers from late spring until the first frosts
Lance-shaped, gray green leaves
Flowers dry wonderfully for winter arrangements

Sow in late spring in prepared, open ground when air temperature does not go below 50°F at night. Germination takes 14–21 days.
OR
Sow in early spring in pots or cell packs using a soil-less seed mix, either peat or peat substitute. Cover with perlite or vermiculite and place under protection at 68°F. Germination takes 7–14 days.

Hypoestes *Acanthaceae*

A genus of tender, mainly evergreen perennials and shrubs with a minimum growing temperature of 50°F. In cooler climates, they make good potted plants because of their attractive foliage. Plant out in the garden in a well-drained soil and in partial shade.

Hypoestes phyllostachya "Pink Splash" (Polka-dot plant)
Medium seeds: 1,620 per ounce
Height 2¹/₂ ft.
Small tubular, lavender flowers
Flowers in summer
Tender evergreen perennial, often grown as an annual in cool climates
Attractive dark green leaves covered with irregular pink blotches

Sow in spring in pots or cell packs using a soil-less seed mix either peat or peat substitute. Cover with perlite or vermiculite, place under protection at 68°F. Germination takes 7–14 days. With good light conditions, this plant can be sown at any time.

Impatiens *Balsaminaceae*

A genus of hardy to tender annuals and evergreen perennials. Plant in soil that is moist but not waterlogged, in partial shade.

Impatiens balsamina (Balsam)
Medium seeds: 280 per ounce
Half-hardy annual
Height 2¹/₂ ft.
Small, cup-shaped, pink or white flowers
Flowers all summer until the first frosts
Lance-shaped, midgreen leaves

Impatiens Novette Series
Small seeds: 2,800 per ounce
Half-hardy evergreen perennial grown as a half-hardy annual
Height 6 in.
Attractive 5-petaled, spurred flowers in red, pink, orange, white, violet, and bicolors
Flowers from summer until the first frosts
Reddish stems with oval, toothed, midgreen leaves

Sow in early spring in pots or cell packs using a soil-less seed mix, either peat or peat substitute. Cover with perlite or vermiculite and place under protection at 70°F. Try to keep the temperature as even as possible. Water with tepid water, not cold water straight from the tap, which would inhibit germination or cause the seedling to wilt. Germination takes 14–21 days.

Laurentia *Campanulaceae*

A genus of tender annuals and perennials. Nearly all of them make lovely potted plants. Plant in a fertile, moisture-retaining soil, not clay or water-logged, in a sunny situation. This plant has been classified as *Isotoma* and *Soleonopsis*, but is currently known as *Laurentia*.
Laurentia axillaris "White" (syn. *Isotoma*) (Flying angels)
Small seeds: 16,800 per ounce

Half-hardy annual
Height 10 in.
Five-petaled, star-shaped, pure white flowers
Flowers all summer
Small, linear, serrated midgreen leaves
Looks stunning in tubs, pots, and baskets.

Sow in late winter in pots or cell packs using a soil-less seed mix, either peat or peat substitute. Cover with perlite or vermiculite and protect at 20°C (68°F). Germination takes 8–14 days.

Recipe for sowing, planting, and growing Laurentia axillaris
This plant is a stunner in a container or hanging basket, flowering profusely all summer. The star-shaped flowers come in lovely shades of pink, white, blue, and pale mauve.

I have seen an old wooden wheelbarrow planted with 3 different shades of flower which cascaded gently over the edge; it looked wonderful.

Ingredients
10 seeds per cell or 15 seeds per pot
1 white 3 x 5 card, folded in half
1 flat with cell packs
OR
1 x 4 in. pot
Soil-less seed mix, either peat or peat substitute
Fine-grade perlite (wetted) or vermiculite
White plastic plant label

Method Fill the cell packs or pot with soil mix, smooth over, tap down and water in well. As the seeds are very fine, it is worth putting a very small amount of seed into the crease of the folded 3 x 5 card. Tap the card gently, in order to sow the seed thinly on the surface of the soil mix. Cover with perlite or vermiculite and label with the plant name and date. Place the flat or pot in a warm, light place out of direct sunlight at an optimum temperature of 68°F. Keep watering to a minimum until germination has taken place, which takes 8–14 days.

Shortly after germination, and when the seedlings have emerged fully, put the containers in a cooler environment 59°F) Plants can be pricked out and potted up 2–3 weeks after germination. They will be ready to plant out approximately 4 weeks after that. If you are using cell packs you can plant out direct into a container or hanging basket as soon as they have rooted right down the cell pack. Keep the container or hanging basket protected, until all threat of frost has passed, then harden off before leaving out all night. If you wish to plant in the garden, make sure the seedling has a period of hardening off, then plant out when there is no threat of frost.

Lavatera *Malvaceae*

A genus of hardy annuals, biennials, and perennials. Plant in well-drained soil and a sunny situation.

Lavatera trimestris Beauty Series
Medium seeds: 420 per ounce
Hardy annual
Height 2 ft.
Shallow, trumpet-shaped flowers in pink, rose, salmon, and white
Flowers from early summer until early autumn
Oval, lobed leaves

Sow in late spring in prepared, open ground when air temperature does not go below 50°F at night. Germination takes 14–21 days.

Limnanthes *Limnanthaceae*

A genus of fully hardy annuals. Plant in fertile, well-drained soil in a sunny position.

Limnanthes douglasii AGM (Meadow foam, Poached-egg flower)
Medium seeds: 364 per ounce
Hardy annual
Height 6 in.
Cup-shaped, lightly scented flowers with yellow centers
Flowers from summer until early autumn in cool climates
Light green, feathery leaves

Sow in late spring in prepared, open ground when air temperature does not go below 50°F at night. Germination takes 14–21 days.

Limonium *Plumbaginaceae*

A genus of fully hardy to frost-tender perennials sometimes grown as annuals. Plant in well-draining soil in a sunny situation.

Limonium sinuatum (*Statice sinuata*) (Annual statice)
Medium seeds: 1,120 per ounce
Half-hardy perennial grown as an annual in cool climates
Height 18 in
Clusters of tiny, papery flowers in blue, pink, or white
Flowers from summer to early autumn
Lance-shaped, lobed, deep-green leaves
Flowers are good for drying

Limonium platyphyllum (*latifolium*) (Sea lavender)
Medium seeds: 2,800 per ounce
Hardy perennial, often grown as an annual
Height 12 in.
Masses of small, blue-mauve clusters of flowers
Flowers throughout the summer
Good for dried-flower arrangements
Large, leathery, dark green leaves

Sow in late spring in prepared open ground when air temperature does not go below 50°F at night. Germination takes 14–21 days.

Lobelia *Campanulaceae*

A large genus of half-hardy annuals, perennials, and subshrubs. The dwarf forms are often used as bedding plants. Plant in a light, free-draining soil with added compost to retain moisture through the summer. Bedding lobelias do not like to dry out, or to be waterlogged. All lobelias hate being wet in winter in cold climates.

The seeds of *Lobelia erinus* are minute. You can buy pelleted seeds from seed suppliers, which are easier to see and use. At the time of publication there are no organic pelleted seeds available.

Lobelia erinus compacta
Minute seeds: 70,000 per ounce
Half-hardy annual
Height 4 in.
Small, 2-lipped flowers in a range of colors—white, purple pink, blue—often with a white eye
Flowers throughout the summer until the first frosts
Small, midgreen oval or lance-shaped leaves
Ideal for containers, tubs, and hanging baskets

Sow in early spring in pots or cell packs using soil-less seed mix, either peat or peat substitute. As these are very fine seeds, mix them with the finest sand or talcum powder for an even sowing. Do not cover. Water from the bottom or with a fine spray and place under protection at 68°F. Germination takes 8–14 days.

Lobelia cardinalis AGM (Cardinal flower) Zones 2–9
Tiny seeds: 56,000 per ounce
Hardy perennial often grown as a hardy annual
Height 3½ ft.
Attractive, 2-lipped scarlet flowers
Flowers from mid- to late summer
Lance-shaped, midgreen or red-bronze leaves
Good for containers and as a summer-flowering border plant

Sow in early spring in pots or cell packs using soil-less seed mix, either peat or peat substitute. As the seeds are very fine, mix them with the finest sand or talcum powder for an even sowing. Do not cover. Water from the bottom or with a fine spray, place in a cold frame under protection at 48°F at night for 10 days. Then give the seeds heat at 68°F. Germination takes 8–14 days after heat is applied.

***Recipe for sowing, planting, and growing* Lobelia cardinalis**
If you want a "Wow!" factor in the garden, then the Cardinal flower is a must. The flowers are vibrant crimson. It is a great plant for the garden, because it likes moist, cool, shady places. It does not mind dry cold in winter, but it hates wet cold. It also makes a stunning feature plant in a container collection. Remember that it likes being moist, so make sure the container does not dry out and position it well out of the midday sun or in partial shade.

Ingredients
10 seeds per cell or 15 seeds per pot (as accurately as you can manage)
Talcum powder or fine plain white flour
1 white card 3 x 5 card, folded in half
1 flat with cell packs

OR

1 x 4 in. diameter pot
Soil-less seed mix, either peat or peat substitute
White plastic plant label

Method Fill the cell packs or pot with soil-less mix, smooth over, tap down and water in well. As the seeds are very fine, it is a good idea to mix them with talcum powder or extrafine white flour, which will show up the seeds . Put a very small amount of this seed mix into the crease of folded 3 x 5, you can then tap the card gently, sowing the seed thinly on the surface of the compost. Do not cover, label with the plant name and date. Place the flat or pot in a cold frame for 10 days, when the temperature does not fall below 48°F at night in the frame. Then bring the flat or pot inside and place in a warm light place out of direct sunlight at an optimum temperature of 68°F. Keep watering to a minimum, until germination has taken place and then water only from the bottom or with a fine spray, in order not to disturb the fine seeds. Germination takes 8–14 days with warmth. Shortly after germination has taken place, and when the seedlings are fully emerged, put the containers in a cooler environment 59°F.

Prick out when the seedlings are large enough to handle or, if you are using cell packs, plant directly in a container or the garden, after a period of hardening off and when there is no threat of frost.

Lunaria *Brassicaceae*

A genus of hardy biennials and perennials that produce lovely, silvery seed pods. Plant in well-drained soil in partial shade.

Lunaria annua (Honesty, Money Plant) Zones 4–8
Medium seeds: 140 per ounce
Biennial
Height 2$\frac{1}{2}$ ft.
Scented, 4-petalled, white or purple flowers
Flowers from spring to midsummer, usually in second year
Rounded, silvery seed pods which are lovely for dried-flower arrangements
Oval, pointed, serrated, midgreen leaves

Sow the seeds fresh in late summer in prepared, open ground. Do not protect the seed from winter weather, as this allows natural stratification to occur (see page 233). The seeds will germinate the following spring and occasionally flower the same year.
OR
Sow fresh seed in autumn in pots using a standard, soil-less seed mix. Cover lightly with the mix and place in a cold frame. Germination will take place the following spring.

Matthiola *Brassicaceae*

A genus of hardy to frost-tender annuals, biennials, perennials, and evergreen subshrubs. Plant in fertile well-drained soil, (prefers lime) in a sun or partial shade.

Matthiola incana (Stock, Brompton stock)
Medium seeds: 1,680 per ounce
Biennial or short-lived perennial

Height 1–2 ft.
Highly scented 4,-petalled spikes of pink, apricot, cream, or lilac flowers
Flowers in summer
Lance-shaped, gray green leaves

Sow in late spring in a prepared open ground when air temperature does not go below 50°F at night. Germination takes 14–21 days.
OR
Sow in early spring in pots or cell packs using soil-less seed mix, either peat or peat substitute. Cover with perlite or vermiculite and place under protection at 65°F. Germination takes 8–14 days.

Mimulus *Scrophulariaceae*

A genus of annuals, perennials, and evergreen shrubs. Plant in a moist to wet soil in a sunny situation.

Mimulus luteus (Monkey flower)
Tiny seeds: 56,000 per ounce
Half-hardy perennial often grown as an annual
Height 12 in.
Yellow, snapdragon-shaped flowers
Flowers throughout the summer
Oval, toothed, hairy midgreen leaves

Mimulus "Tigrinus"
Minute seeds: 64,400 per ounce
Half-hardy annual
Height 8 in.
Lovely yellow flowers, splashed with red
Flowers throughout the summer
Oval, toothed, midgreen leaves

Sow in early spring in pots or cell packs using soil-less seed mix, either peat or peat substitute. Cover with perlite or vermiculite, place under protection at 65°F. Germination takes 14–21 days.

Mirabilis *Nyctaginaceae*

A genus of annuals and perennials. Plant in well-drained soil and a sheltered position in full sun.

Mirabilis jalapa (Four o'clock, Marvel of Peru)
Large seeds: 28 per ounce
Tender perennial often grown as an annual
Height 2–4 ft.
Fragrant trumpets of crimson, pink, white, or yellow flowers that open in the evening (after 4 o'clock)
Flowers from summer until frost
Mid-green, oval, tipped leaves

Sow in early spring in pots or cell packs using soil-less seed mix, either peat or peat substitute. Cover with perlite or vermiculite, place under protection at 65°F. Germination takes 14–21 days.

Papaver somniferum

Papaver *Papaveraceae*

A genus of hardy annuals, biennials, and perennials. A stunning variety is *Papaver somniferum* "Peony-flowered mixed". Double varieties are available in single and mixed colors. Plant in a moist but well-drained soil in sun or partial shade.

Papaver somniferum (Opium poppy)
Small seeds: 8,400 per ounce
Hardy annual
Height 2¹/₂ ft.
Large, single flowers in shades of red, pink, or purple, with darker centers
Flowers in early summer
Oblong, gray green, lobed leaves

Papaver nudicaule "Meadow Pastels" (Iceland poppy, Arctic poppy)
Zones 2–6
Small seeds: 1,120 per ounce
Hardy perennial, often grown as an annual, especially in warmer climates
Fragrant flowers in white, yellow, orange, and red
Flowers in summer
Soft, green, oval, toothed leaves

Sow fresh seeds in late summer in prepared, open ground. Do not cover. Germination will take place the following spring.
OR
Sow the seeds in spring in prepared, open ground when the air temperature does not go below 48°F at night. Do not cover the seed. Germination takes 2–4 weeks.

Petunia *Solanaceae*

Half-hardy perennials usually grown as annuals. Plant in a fertile, well-drained soil in a sunny sheltered position.

Petunia x *hybrida* Grandiflora types
Small seeds: 25,200 per ounce
Half-hardy perennial
Height 6–12 in.
Flared, trumpet-shaped flowers available in many colors, including bicolors
Flowers throughout summer until the first frosts
Oval, midgreen leaves

Petunia x *hybrida* Multiflora types
Small seeds: 22,400 per ounce
Half-hardy perennial grown as an annual
Height 6–12 in.
Flared, trumpet-shaped flowers, smaller than Grandiflora types
Flowers are often double frilled
Available in many colors, including bicolors
Flowers throughout the summer until the first frosts
Oval, midgreen leaves

Sow in early spring in pots or cell packs using a standard, soil-less seed mix, either peat or a peat substitute. Cover with perlite or vermiculite and place under protection at 70°F. Germination takes 7–10 days.

Phacelia *Hydrophyllaceae*

A genus of hardy annuals, biennials, and perennials, good for the front of borders or in containers. Plant in fertile, well-drained soil in a sunny situation.

Phacelia campanularia (California bluebell)
Medium seeds: 2,800 per ounce
Hardy annual
Height 8 in.
Bell-shaped, gentian-blue flowers
Flowers in summer
Oval, serrated, midgreen leaves

Sow fresh seeds in autumn in prepared open ground. Do not cover the seed. Germination will take place the following spring.
OR
Sow the seeds in spring in prepared open ground, when air temperature does not go below 48°F at night. Do not cover the seed. Germination takes 2–4 weeks.

Phlox *Polemoniaceae*

A genus of annuals, perennials, and evergreens. Plant in a fertile, moist, not waterlogged soil in sun or semishade.

Phlox drummondii Twinkle Series
Medium seeds: 1,400 per ounce
Half-hardy annual
Height 6 in.
Star-shaped flowers in many shades of bright colors, some with contrasting centers
Flowers all summer
Pale-green, lance-shaped leaves

Sow the seeds in spring in prepared, open ground, when air temperature does not go below 48°F at night. Do not cover the seed. Germination takes 2–4 weeks.
OR
Sow in early spring in pots or cell packs using a standard soil-less seed mix, either peat or a peat substitute. Do not cover. Place under protection at 60°F. Germination takes 10–20 days. Phlox does not perform well if transplanted from a seed tray, so use cell packs and pots.,

Recipe for sowing, planting, and growing* Phlox drummondii *Twinkle Series

This is a very popular dwarf bedding annual, which produces a profusion of lovely starry flowers in a mixture of colors—pinks, purples, reds—with contrasting centers. For example, pale pink can have a darker pink center, or purple can have white edges and white centers. From on high, you could describe these flowers as paint splodges. They look great in the front of the border or in containers and hanging baskets.

Ingredients
4 seeds per cell or 8 seeds per pot
1 flat with cell packs
OR
1 x 4 in. pot
Soil-less seed mix, either peat or peat substitute
White plastic plant label

Method Fill the cell packs or pot with spoil-less mix, smooth over, tap down and water in well. Sow thinly in the container, spacing the seeds well. Press gently into the soil-less mix with the palm of the hand. Do not cover, label with the plant name and date. Place the flat or pot in a warm, light place out of direct sunlight, at an optimum temperature of 60°F. Keep watering to a minimum until germination has taken place, which takes 10–20 days in spring with warmth. Pot up or prick out approximately 4 weeks later. If you are using cell packs, you can plant directly in containers. Wait to plant out in the garden until the young plants have had a period of hardening off and there is no threat of frost.
OR
Sow directly in a prepared site in the garden, when the night time temperature does not fall below 48°F at night. Space the seeds 4 in. apart, press gently into the soil, do not cover, water in well. If you are plagued by birds or mice it might be worth covering with floating seed covers to stop your seeds being stolen. Germination takes 14–28 days. Remove the cover as soon as you see that they have germinated.

Portulaca *Portulacaceae*

A genus of annuals and perennials with succulent leaves and flowers that open in the sun and close in the shade.

Portulaca grandiflora (Moss rose)
Small seeds: 25,200 per ounce
Half-hardy annual
Height 6–8 in.
Bowl-shaped flowers in shades of yellow, pink, red, and white, with conspicuous stamens
Flowers from summer until frost
Bright green, lance-shaped, succulent leaves

Sow in early spring in pots or cell packs using a standard, soil-less seed mix, either peat or a peat substitute. Cover with perlite or vermiculite, place under protection at 68°F. Germination takes 10–14 days.

Ricinus *Euphorbiaceae*

A genus of a single species, half-hardy perennial grown as an annual in cooler climates. Plant in a humus-rich, well-drained soil in a sheltered sunny position.

Ricinus communis "Gibsonii" (Castor oil plant, Castor bean)
ALL PARTS OF THIS PLANT ARE POISONOUS
Large seeds: 5–6 per ounce
Half-hardy perennial
Height 6–8 ft.
Green, petal-free, female flowers are followed by red capsules, each containing 3 seeds
Flowers from summer until frost
Dark green stems with large, lobed, toothed, bronze leaves

Sow in early spring in pots or cell packs using a standard, soil-less seed mix, either peat or a peat substitute. Cover with perlite or vermiculite, place under protection at 70°F. Germination takes 14–21 days.

Rudbeckia *Asteraceae*

A genus of annuals, biennials, and perennials. Plant in moist but well-drained soil in sun or partial shade.

Rudbeckia hirta "Rustic Dwarfs" (Gloriosa dailsy)
Small seeds: 5,600 per ounce
Short-lived perennial grown as a hardy annual
Height 1–3¹/₂ ft.
Large, daisy-like flowers, yellow, mahogany, or bronze, with dark centers
Flowers all summer until the first frosts
Mid-green, lance-shaped leaves

Sow in early spring in pots or cell packs using a standard, soil-less seed mix, either peat or a peat substitute. Cover with perlite or vermiculite and place under protection at 65°F. Germination takes 10–14 days.

Salpiglossis *Solanaceae*

Half-hardy annuals and biennials. Plant in a rich, well-drained soil in a sunny situation. Flower stems might need support in exposed sites.

Salpiglossis sinuata Bolero Series (Painted tongue)
Small seeds: 11,200 per ounce
Half-hardy annual
Height 2 ft.
Trumpet shaped, veined flowers in a mixture of colors, yellow, orange, red, and blue
Flowers in summer
Pale-green, lance-shaped leaves

Sow in early spring in pots or cell packs using a standard, soil-less seed mix, either peat or a peat substitute. Cover the seed with black plastic, check regularly, remove as soon as germination starts. Place under protection at 65°F. Germination takes 10–14 days.

Salvia *Lamiaceae*

see also PERENNIALS and HERBS. A large genus of annuals, biennials, perennials, and evergreen shrubs. Plant in well-drained, fertile soil in a sunny situation.

Salvia splendens "Blaze of Fire"
Medium seeds: 840 per ounce
Half-hardy perennials often grown as an annual
Height 12 in.
Dense spikes of brilliant, scarlet, 2-lipped flowers
Flowers all summer until the first frosts
Bright green, oval, serrated leaves

Sow in early spring in pots or cell packs using a standard, soil-less seed mix, either peat or a peat substitute. Cover with perlite or vermiculite, place under protection at 65°F. Germination takes 14–21 days.

Sanvitalia *Asteraceae*

A genus of hardy perennials and annuals. Plant in fertile, well-drained soil in a sunny situation.

Sanvitalia procumbens "Gold Braid" (Creeping zinnia)
Small seeds: 4,480 per ounce
Hardy annual
Height 6 in.
Daisy-like yellow flowers with black centers
Flowers all summer
Mid-green, oval, pointed leaves

Sow in early spring in pots or cell packs using a standard, soil-less seed mix, either peat or a peat substitute. Cover with perlite or vermiculite and place under protection at 59°F. Germination takes 14–21 days.

aquatic plants

Water is wonderful in a garden. There is nothing more beautiful than the sun dancing over the water, and the plants growing in or beside it reflected in the shimmering mirror. Even a small pond will give you great pleasure and will attract lots of wildlife to your garden. Unlike any other plants in this book, water plants must be kept moist at all times during germination. Approximate seed sizes only are given in this chapter as more specific information is not easily available. Refer to page 242 for more information on seed sizes.

Alisma *Alismataceae*

A genus of hardy, deciduous, perennial, marginal water plants. Plant in muddy, wet soil or in water up to 8in deep in a sunny position.

Alisma plantago-aquatica (Water plantain) Zones 5–8
Small seeds
Hardy deciduous perennial
Height up to 2^1/$_2$ ft.
Loose clusters of small pale pink/white flowers
Flowers in summer
Oval, bright green leaves

Sow seeds in autumn in pots using standard loam-based seed mix, mixed with coarse grit to a ratio of 1 part soil mix + 1 part grit. Cover with coarse horticultural sand and stand in water so that the soil mix remains wet. Place outside, exposed to all weathers (See "Breaking Seed Dormancy", page 233, for more information.) Germination takes 4–6 months. Overwinter young plants in a cold frame and keep damp.

Caltha *Ranunculaceae*

A genus of hardy, deciduous, perennial, marginal water plants, bog plants, or alpine plants, grown for their attractive flowers. Plant in a moist soil in a sunny position.

Caltha palustris AGM (Kingcup, Marsh marigold) Zones 3–9
Small seeds
Hardy deciduous perennial
Height 2 ft.
Clusters of cup-shaped, bright golden-yellow flowers
Flowers in spring
Rounded dark green leaves

Sow seeds in late summer in pots or cell packs, using standard soil-less seed mix, either peat or peat substitute. Cover with coarse horticultural sand. Place under protection at 50°F. During germination, use a hand spray to keep the mix moist. Germination takes 4–6 weeks. Overwinter young plants in a cold frame; keep damp.

Gunnera *Gunneraceae*

A genus of hardy perennials, which generally have very large leaves. Plant in poor, moist soil in a sheltered, sunny position.

Gunnera manicata AGM Zones 7–10
Medium seeds
Hardy perennial
Height 8 ft.
Light green flower spikes followed by orange-brown seed pods
Flowers in early summer
Very large, rounded, serrated green leaves

Sow fresh seeds in late summer, as soon as they are ripe, in pots or cell packs using standard soil-less seed mix, either peat or peat substitute. Cover with coarse horticultural sand. Place under protection at 70°F, using a hand spray to keep the mix moist but not wet. Germination takes 4–6 weeks. Overwinter young plants in a cold frame; keep damp.

Hottonia *Primulaceae*

A genus of hardy, deciduous, perennial, submerged water plants. Plant in still or running water in a sunny situation.

Hottonia palustris (Water violet) Zones 5–9
Medium seeds
Hardy perennial
Height 12 in.
Delicate, pale mauve flowers with a yellow throat
Flowers in summer
Finely divided, midgreen leaves

Sow fresh seeds in late summer in pots using standard soil-less seed mix, either peat or peat substitute. Cover with coarse horticultural sand. When you have sown the seed, stand the pots in water so the mix remains wet at all times. Place under protection at 60°F. Germination takes 4–6 weeks. Overwinter young plants in a cold frame and keep damp.

***Recipe for sowing, planting, and growing* Hottonia palustris**
This is a lovely and elegant flower that should have a place in every pond. It is not a violet, as its common name suggests, it is a member of the primrose family and, as such, is the only one to grow in water.

Ingredients
10 seeds per pot
1 x 4 in. pot
Standard soil-less seed mix, either peat or peat substitute
White plastic plant label
Container without drainage holes, large enough to hold the pot

Method In spring fill the pot with soil-less mix, smooth over, tap down and water in well. Sow the fresh seeds thinly on the surface of the mix. Press gently in with the palm of the hand. Cover thinly with soil-less mix, and label with the name and date. Place pot in the container, fill the container with water to just below the rim of the pot and place in a warm, light place out of direct sunlight, at an optimum temperature of 60°F. Germination will take anything from 2–6 weeks. Once germinated, remove from the heat, keep the two containers together and place in a warm, light place to allow the seedlings to grow on. When large enough to handle, pot up into soil-less potting mix. Place the new pots in containers holding water. Grow on in a cold frame for the first winter, and plant in the pond the following spring.

Hydrocharis *Hydrocharitaceae*

A genus with a single, hardy species. Plant in still water in a sunny position.

Hydrocharis morsus-ranae (Frogbit) Zones 7–10
Small seeds
Hardy deciduous perennial
Height 4 in.
Small white flowers with a yellow eye
Flowers in summer
Rosettes of kidney-shaped olive green leaves

Caltha palustris

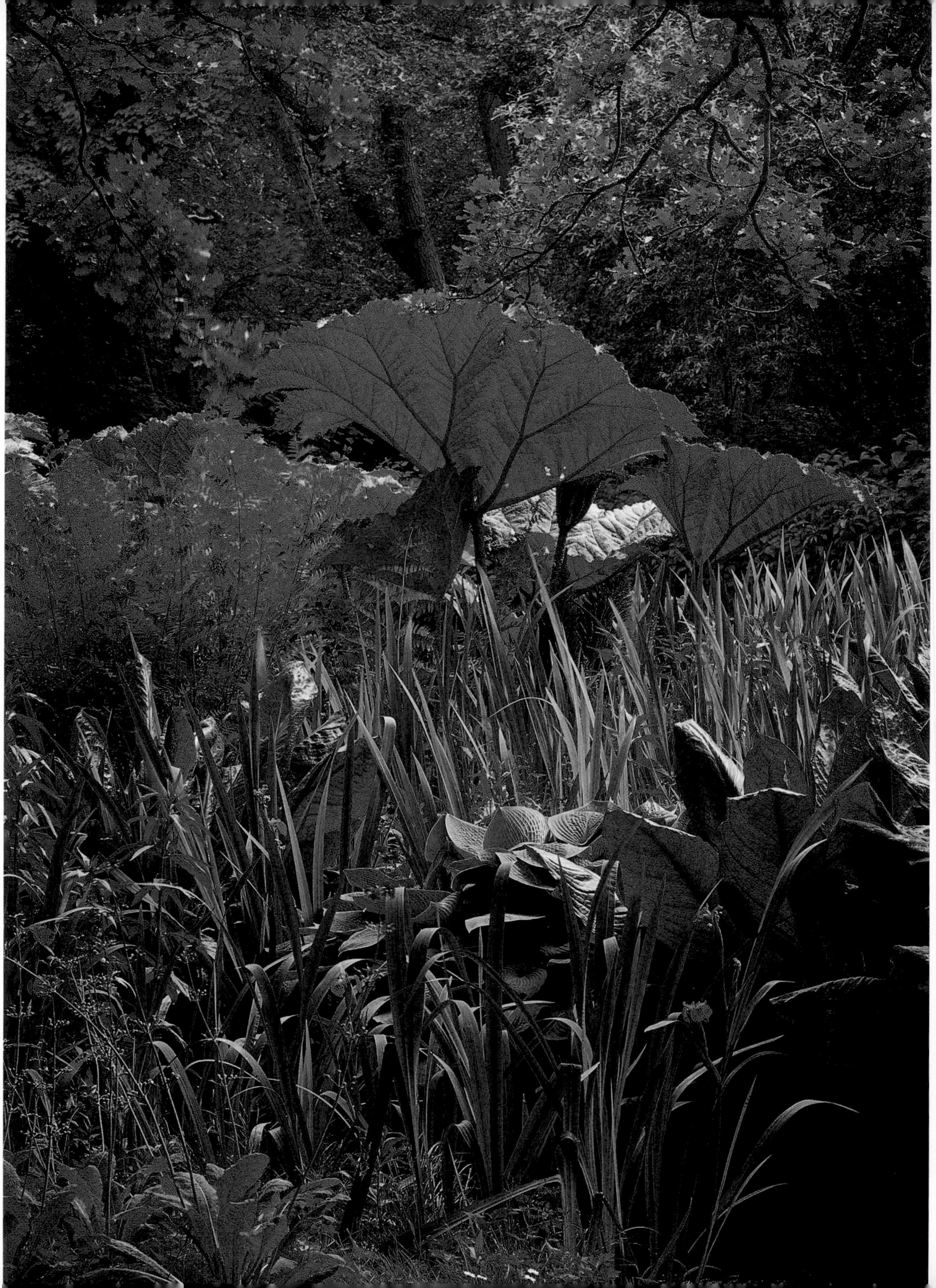

Gunnera manicata

Sow seeds in late summer in pots, using standard soil-less seed mix, either peat or peat substitute. Cover with coarse horticultural sand. When you have sown the seed, stand the pots in water, so that the mix remains wet at all times. Place under protection at 50°F. Germination takes 4–6 weeks. Overwinter young plants in a cold frame.

Menyanthes *Menyanthaceae*

A genus of hardy perennial, deciduous, marginal water plants. Plant in a moist soil in a sunny position.

Menyanthes trifoliata (Bog Bean) Zones 4–8
Small seeds
Height up to 12 in.
Hardy perennial
Lovely pink buds, which, when open, are attractive white stars
Flowers in spring
Oval green leaves

Sow seeds in spring in pots, using standard soil-less seed mix, either peat or peat substitute. Press the seeds well into the wet soil-less mix, do not cover. Stand the pots in water so the mix remains wet at all times. Place the container in a cold frame or cold greenhouse. Germination takes 4–6 weeks. Grow on in a cold frame for the first winter, plant in the pond the following spring.

Nymphaea *Nymphaeaceae*

A genus of tender and hardy perennial water plants grown for their lovely flowers. Plant in still water in a sunny position.

Nymphaea alba (Waterlily) Zones 3–10
Small seeds
Hardy deciduous perennial
Height 4 in.
Lovely, star-shaped, semidouble, white flowers with golden centers
Flowers in summer
Round dark-green floating leaves

Waterlily seeds germinate under water. In summer, get into the pond and select some fading flowers. Wrap muslin around the flower head. When the flower has gone, and the seed pod has sunk below the surface of the water, it is time to pick it. Remember, the pods explode when they are ripe.

Sow seeds in spring in pots using standard soil-less seed mix, either peat or peat substitute. Push the seeds well into the soil-less mix and submerge in water. Place the container in a cold frame or a cold greenhouse. Germination takes 4–6 weeks. When large enough, plant back in the pond.

Ranunculus *Ranunculaceae*

A genus of annuals, perennials, and aquatics, some of which are evergreen. Plant in any soil that does not dry out in summer, in sun or partial shade.

Ranunculus aquatilis (Water buttercup, Water-crowfoot)
Zones 4–9
Small seeds
Hardy perennial
Height 12 in.
Lovely, small, white cupped flowers
Flowers in summer
Small, midgreen, wedge-shaped leaves that can be very divided

Sow seeds in autumn in pots using standard soil-less seed mix, either peat or peat substitute. Press the seeds well into wet soil-less mix, do not cover. Stand the pots in water, so that the mix remains wet at all times. Place the container in a cold frame or cold greenhouse. Germination takes 4–6 weeks. Overwinter in the cold frame, keep moist at all times, plant out in the spring.

Rodgersia *Saxifragaceae*

A genus of hardy rhizomatous perennials. Plant in a moist soil in partial shade.

Rodgersia aesculifolia AGM Zones 5–6
Small seeds
Hardy perennial
Height 3 1/2 ft.
Clusters of fragrant, pinky-white flowers
Flowers in summer
Bronze, crinkled lobed foliage, rather like the horse chestnut
Ideal for growing around ponds or in a bog garden

Sow seeds in spring in pots using standard, loam-based seed mix. mixed with coarse grit. Mix in a ratio of 1 part soil-less mix + 1 part grit. Put a layer of moss on the surface of the soil-less mix and sow the seeds directly on the moss. Water in, do not cover. When you have sown the seeds, stand the pots in water so that the mix remains wet at all times. Place under protection at 60°F. Germination takes 2–3 weeks. Grow on in a cold frame for the first winter, plant the following spring.

Sagittaria *Alismataceae*

A genus of tender to fully hardy perennial, deciduous, marginal, and submerged water plants. Plant in water in a sunny position.

Sagittaria sagittifolia (Common arrow-head) Zones 4–9
Small seeds
Hardy deciduous perennial
Height 18 in.
White flowers with dark purple centers
Flowers in summer
Mid-green, arrow-shaped leaves

Sow seeds in late summer in pots using standard, loam-based seed mix, mixed with coarse grit. Mix to a ratio of 1 part soil mix + 1 part grit. Cover with coarse horticultural sand then stand the pots in water, so that the mix remains wet at all times. Place under protection at 50°F. Germination takes 2–3 weeks. Grow on in a cold frame for the first winter, plant in the pond the following spring.

cacti & succulents

Cacti and succulent plants are immensely collectable and are, in the majority of cases, very straightforward to grow from seed. As I grow only a few cacti and a couple of succulents on the farm, I consulted Bryan Goodey, one of the U.K.'s foremost cacti growers, before writing this chapter. I am very grateful for his help. Seed is usually sold by amount rather than weight, so, in this section, I give an indication of seed size only. For the best results, seed should be sown fresh, and so seed collected at home can prove more reliable than purchased seed. All cacti need as much light as possible when germinating, so do not cover. If you have "grow lights" use them for the first twelve weeks, as this will be of great benefit. Even a household light bulb in an flexxible-arm lamp will help.

Astrophytum myriostigma, see page 69

climbers

Climbing plants are very useful in the garden. They can transform a tree stump into a living feature, cascade over a wall, climb up and cover a boring fence, or even camouflage a garage. Some climbers have the most attractive foliage, while other climbers have stunning flowers, and some varieties combine the two. Whichever climber you choose, it is sure to be a good investment, whatever the size of your garden.

Actinidia *Actinidiaceae*

A genus of hardy to half-hardy climbers. Plant in any well-drained soil that does not dry out in summer, and in full sun to partial shade. To obtain fruits, grow both male and female plants.

Actinidia chinensis (Kiwi) Zones 8–9
Medium seeds: 1,400 per ounce
Hardy deciduous perennial
Height up to 30 ft.
Clusters of cup-shaped, white flowers, which turn cream colored as they age, followed by edible, hairy, brown fruits
Flowers in summer
Large, heart-shaped leaves

Sow fresh seeds extracted from the fruit in autumn in pots or cell packs using a standard, soil-less seed mix. Cover with perlite or vermiculite, place under protection at 68°F. Germination takes 2–3 weeks. Overwinter young plants in a frost-free environment.
OR
Sow old seed in autumn in pots or cell packs, using standard, loam-based seed mix, mixed with coarse horticultural sand. Mix in a ratio of 1 part soil mix + 1 part sand. Cover with coarse horticultural sand. Place outside, exposed to all weathers (see "Breaking Seed Dormancy", page 233, for more information). Germination takes 4–6 months but can be erratic. Be patient and do not discard the container.

Cardiospermum *Sapindaceae*

A genus of half-hardy herbaceous or deciduous climbers grown for their fruits. Plant in a light, well-drained soil in a partly shaded or sunny position.

Cardiospermum halicacabum (Balloon vine, Love-in-a-Puff)
Medium seeds: 39 per ounce
Tender perennial grown as an annual
Height up to 10 ft.
Whitish flowers followed by grass-like straw-colored, heart-shaped seed pods
Flowers in summer
Leaves divided into heavily toothed leaflets

Sow seeds in early spring in pots or cell packs, using a standard, soil-less seed mix. Cover with perlite or vermiculite, place under protection at 68°F. Germination takes 3–4 weeks. If growing as a container plant, use a soil-less mix, either peat or a peat substitute.

Clematis *Ranunculaceae*

A genus of fully- to half-hardy deciduous and evergreen climbers and herbaceous perennials. Plant in a rich, well-drained soil in a sunny position; however, will tolerate some shade.

Clematis tangutica (Virgin's bower) Zones 5–9
Medium seeds: 1,680 per ounce
Hardy perennial

Height up to 20 ft.
Lantern-shaped, single, bright yellow flowers, followed by feathery seed heads
Flowers in summer and early autumn
Pointed, oval, green leaves

Sow seeds in early spring in pots or cell packs, using a standard, soil-less seed mix. Cover with perlite or vermiculite. Place under protection at 68°F for 2 weeks, then put in a cold frame at 36–40°F for a month. Move container to a warm situation, increasing the warmth slowly until the seed germinates, about 1 week after being introduced to the warmth.

Clematis viticella Zones 4–9
Medium seeds: 112 per ounce
Hardy perennial
Height up to 10 ft.
Open, bell-shaped, single, purple mauve flowers
Flowers in late summer
Oval leaves divided into lance-shaped leaflets

Sow seeds in early spring in pots or cell packs, using a standard, soil-less seed mix. Cover with perlite or vermiculite, place under protection at 68°F. Germination takes 7–14 days.

Cobaea *Cobaeaceae*

A genus of a single species, half-hardy, evergreen climber. Plant in any well-drained soil in a sunny position.

Cobaea scandens (Cathedral bells, Cup and saucer vine)
Medium seeds: 42 per ounce
Tender perennial, often grown as an annual
Height 12–15 ft.
Bell-shaped, green/white flowers which turn deep purple with age
Flowers from late summer until the first frosts
Leaves divided into oval leaflets

Sow seeds in early spring in pots or cell packs, using a standard, soil-less seed mix. Cover with perlite or vermiculite. Place under protection at 70°F. Germination takes 4–10 days.

Convolvulus *Convolvulaceae*

A genus of climbing annuals, perennials, and evergreen shrubs. Plant in any well-drained soil in a sunny position.

Convolvulus tricolor (*Convolvulus minor*)
Medium seeds: 140 per ounce
Tender annual
Height up to 2 ft.
Saucer-shaped, blue or white flowers with yellowish-white throats
Flowers in summer
Oval, lance-shaped, midgreen leaves

Sow seeds in early spring in pots or cell packs, using a standard, soil-less seed mix. Cover with perlite or vermiculite. Place under protection at 65°F. Germination takes 3–4 weeks.

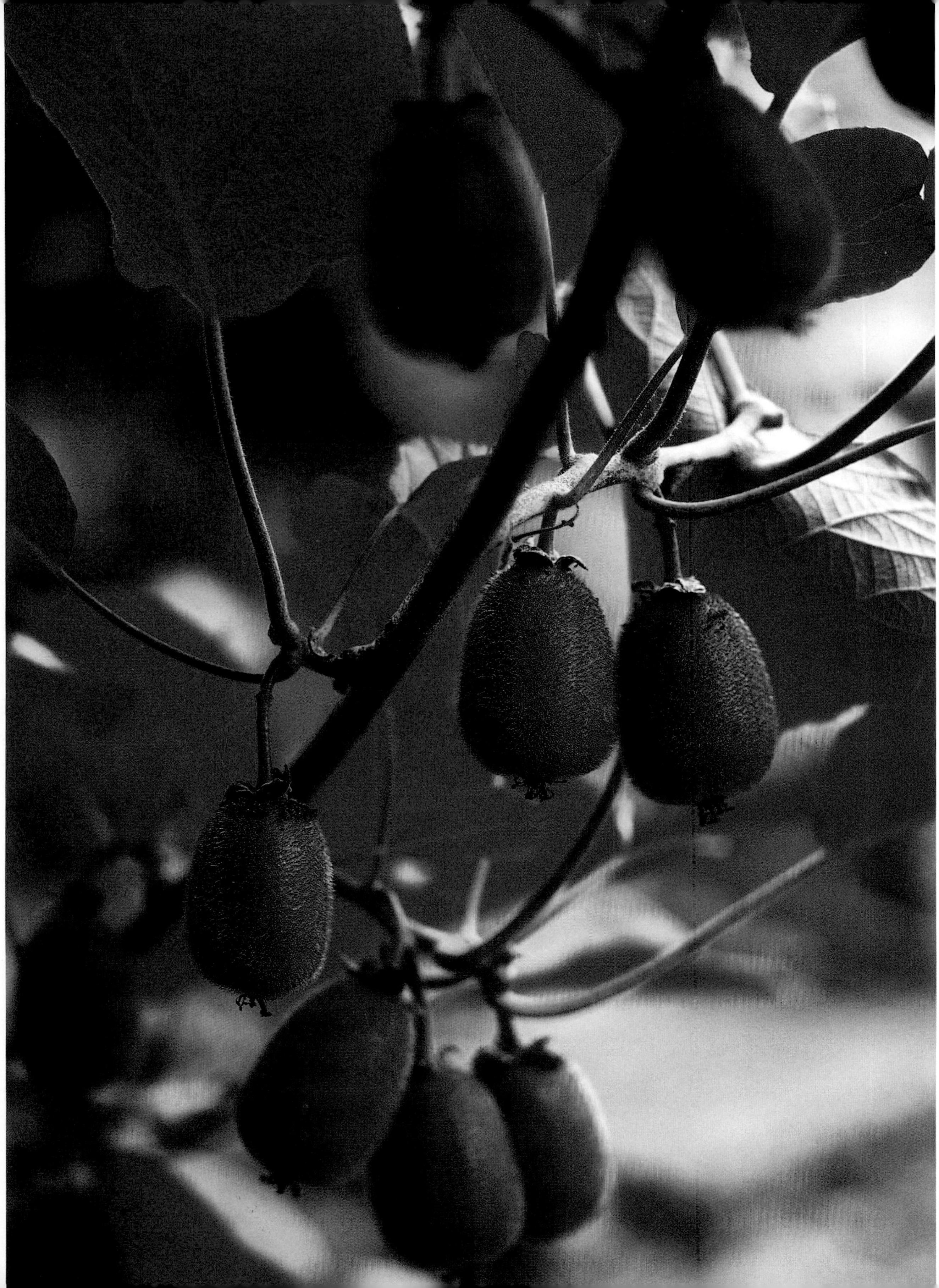

Actinidia chinensis

Dolichos *Papilionaceae*

A genus of a single species, half-hardy, perennial climber grown for its attractive pea-like flowers. Plant in any well-drained soil in a sunny position.

Dolichos lablab (*Lablab purpureus*) (Hyacinth bean)
Large seeds: 14–28 per ounce
Tender perennial, often grown as an annual
Height up to 30 ft.
Purple, pink or white flowers, followed by long pods with
    edible seeds
Flowers from summer until frost
Purple-hued stems, leaves and seed pods

Sow seeds in early spring in pots or cell packs, using a standard, soil-less seed mix. Cover with perlite or vermiculite, place under protection at 70°F. Germination takes 5–8 days.

Eccremocarpus *Bignoniaceae*

A genus of half-hardy, evergreen climbers. Plant in any well-drained soil in a sunny position.

Eccremacarpus scaber (Glory vine)
Large seeds: 28 per ounce
Tender perennial, often grown as an annual
Height up to 10 ft
Clusters of small, orange flowers that are followed by inflated
    fruit pods
Flowers in summer
Small, midgreen leaves

Sow seeds in early spring in pots or cell packs, using a standard, soil-less seed mix. Cover with perlite or vermiculite and place under protection at 65°F. Germination takes 3–4 weeks.

Hardenbergia *Papilionaceae*

A genus of half-hardy, evergreen climbers and subshrubs. Plant in well-drained soil that does not dry out in summer, prefers a sunny position.

Hardenbergia comptoniana
Medium seeds: 224 per ounce
Evergreen, tender perennial. Minimum temperature 45°F
Height up to 8ft
Racemes of pea-like, deep purple-blue flowers
Flowers in early summer
Narrow oval leaves

Soak the seeds in warm water for 24 hours before sowing. Sow seeds in early spring in pots or cell packs, using a standard, soil-less seed mix. Cover with perlite or vermiculite, place under protection at 68°F. Germination takes 6–10 days.

Hedera *Araliaceae*

A genus of hardy and half-hardy, evergreen, perennial climbers and trailing plants. Plant in well-drained soil in semi- to full shade.

Hedera helix (Ivy, English ivy) TOXIC PLANT Zones 4–9
Medium seeds: 56 per ounce
Hardy, evergreen perennial
Height up to 100 ft.
Yellowish-green flowers, followed by round black berries
Flowers in late summer
Five-lobed, dark-green, glossy leaves

Sow seeds extracted from berries in autumn in pots or cell packs, using standard, loam-based seed mix, mixed with coarse horticultural sand. Mix in a ratio of 1 part soil mix + 1 part sand. Cover with coarse horticultural sand. Place outside, exposed to all weathers (see "Breaking Seed Dormancy", page 233, for more information). Germination takes 4–6 months but can be erratic, so be patient and do not discard container.

Recipe for sowing, planting, and growing **Hedera helix**

Ivies are lovely plants. As evergreens, they give interest all year and are very important for wildlife. The late flowers are full of nectar and the berries are eaten by the birds well into the winter.

Ingredients
Rubber gloves
Kitchen towel
3 seeds per cell or 7 seeds per pot
1 flat with cell packs
OR
1 x 4 in. pot
Standard, loam-based seed mix, mixed with coarse horticultural sand.
    Mix in a ratio of 1 part soil mix + 1 part sand for autumn sowing
Extra coarse sand
White plastic plant labels

Method Collect ivy seeds in winter. They are ripe when soft to touch. Wear rubber gloves because ivy is a toxic plant. To extract the seed from the berry, which is a pithy variety, squeeze the fruit gently between your fingers and the seed will pop out. Do not allow the seeds to dry out. Either sow immediately or store in damp sand. Fill the cell pack or pot with soil mix, smooth over, tap down and water in well. Sow thinly on the surface of the mix, press the seeds gently into the mix and cover with coarse sand. Label with the plant name and date. Place the pot outside, on a level surface, so that it is exposed to all weathers, including frosts. Do not worry if you live in a snowy area and the containers get immersed in snow, because melting snow will aid germination.

If you do not live in an area that will get a winter frost, it is a good idea to put the seed mixed with a handful of damp sand in a clearly labelled, sealed plastic bag and place in a refrigerator for 3 weeks. As a precaution, because the seed is toxic, first place the sealed bag in another bag, seal and mark clearly. Remove after 3 weeks and sow as mentioned above, then place outside. Whichever method you use, germination is a bit erratic, taking anything from 1–12 months. Do not give up and discard your soil mix, it might just germinate next week!

Humulus *Cannabaceae*

A genus of hardy and half-hardy herbaceous climbers. Plant in any well-drained soil in sun or semishade.

Humulus lupulus (Hops) Zones 5–8
Medium seeds: 112 per ounce
Hardy perennial
Height up to 20 ft.
Clusters of female, greenish pendant flowers
Flowers in summer
Green leaves divided into 3 or 5 lobes

Sow seeds in autumn in pots or cell packs, using standard, loam-based seed mix, mixed with coarse horticultural sand. Mix in a ratio of 1 part soil mix + 1 part sand. Cover with coarse horticultural sand. Place outside exposed to all weathers (see "Breaking Seed Dormancy", page 233, for more information). Germination takes 4–6 months but can be erratic. Be patient and do not discard container, as the seeds can germinate in the following year.

Ipomoea *Convolvulaceae*

A genus of half-hardy perennials, biennials, and evergreen shrubs or climbers. Plant in humus-rich, well-drained soil in a sunny position.

Ipomoea purpurea "Grandpa Otts"
Medium seeds: 70 per ounce
Semievergreen, tender perennial, often grown as an annual
Height up to 10 ft.
Profusion of intense violet-blue flowers with pink throats
Flowers from summer until frost
Green, heart-shaped leaves

Soak the seeds overnight. Sow in early spring in pots or cell packs, using a standard, soil-less seed mix. Cover with perlite or vermiculite and place under protection at 68°F. Germination takes 6–10 days.

Ipomoea lobata (*Mina lobata*) (Spanish flag, Firecracjer vine)
Medium seeds: 84 per ounce
Semievergreen, tender perennial, often grown as an annual
Height up to 15 ft.
One-sided clusters of pretty, small, tubular, 2-toned flowers of red and cream
Flowers from late summer until frost
Three-lobed, green leaves

Sow seeds in early spring in pots or cell packs, using a standard, soil-less seed mix. Cover with perlite or vermiculite and place under protection at 70°F. Germination takes 3–6 days.

Humulus lupulus

Lathyrus latifolius

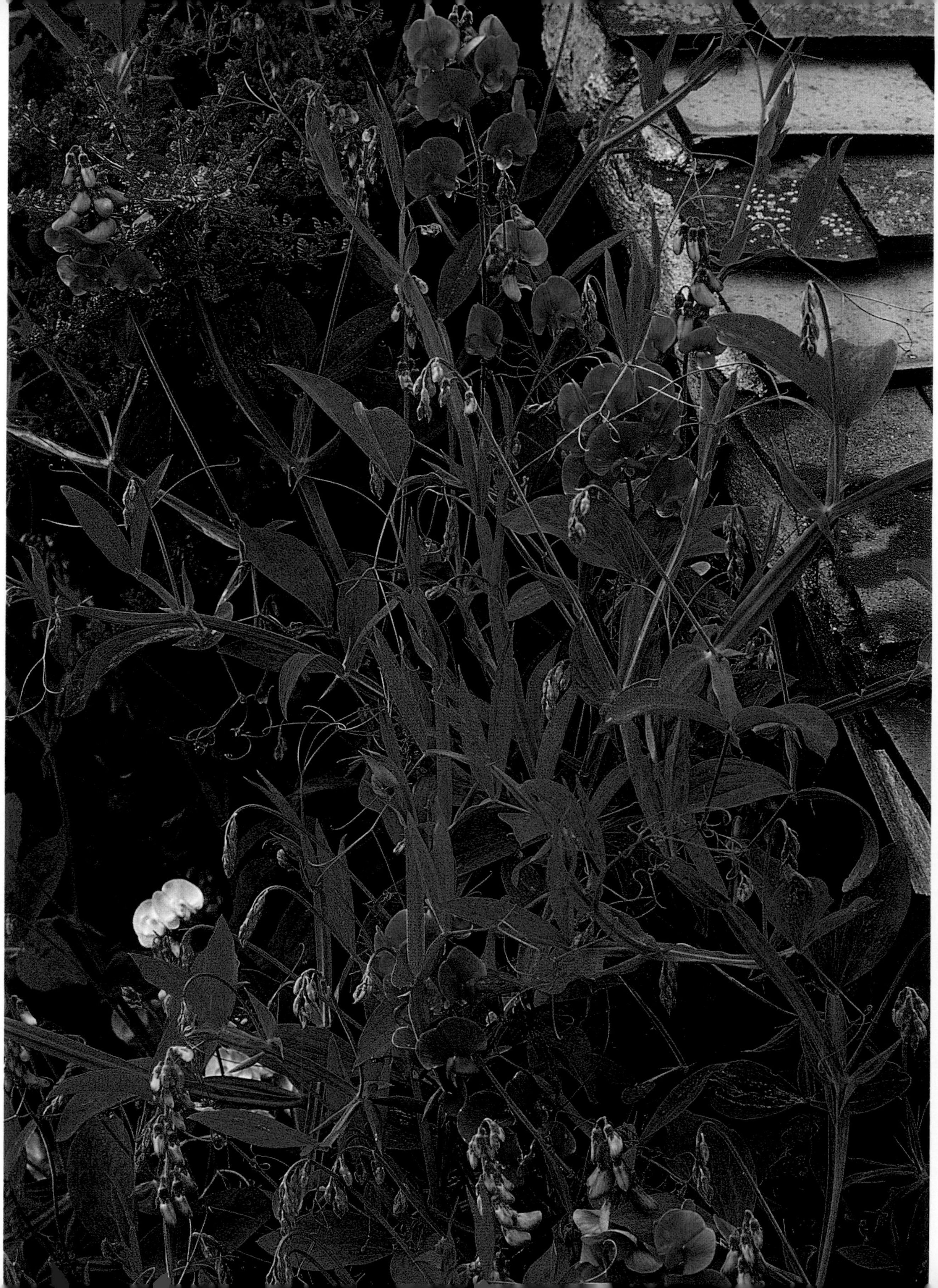

Jasminium *Oleaceae*

A genus of hardy and half-hardy, deciduous or evergreen shrubs and climbers. Plant in fertile, well-drained soil in a sunny position.

Jasminum officinale (Common jasmine) Zones 9–19
Medium seeds: 56 per ounce
Semievergreen, hardy perennial
Height up to 40 ft.
Clusters of pink buds, which develop into scented, small, white flowers followed by black berries
Flowers in summer
Green leaves divided into small pointed leaflets

Sow seeds extracted from berries in autumn in pots or cell packs, using standard, loam-based seed mix, mixed with coarse horticultural sand. Mix in a ratio of 1 part soil mix + 1 part sand. Cover with coarse horticultural sand, then place outside, exposed to all weathers (see "Breaking Seed Dormancy", page 233, for more information). Germination takes 4–6 months but can be erratic, so be patient and do not discard the container.

Lathyrus *Papilionaceae*

A genus of tendril climbers, which all have attractive and, in the case of *Lathyrus odoratus*, sweet-smelling flowers. Plant in well-draining soil in full light. To maintain flowering, dead head regularly.

Lathyrus latifolius (Everlasting pea) Zones 4–9
Medium seeds: 56 per ounce
Hardy herbaceous perennial
Small racemes of pink/purple flowers
Flowers in summer
Broad, lance-shaped leaves with a pair of leaflets

Lathyrus odoratus (Sweet pea)
Medium seeds: 36 per ounce
Hardy annual
Height up to 3m (10ft)
Scented flowers in shades of pink, white, blue, purple, and apricot
Flowers in early summer
Oval, midgreen leaves with tendrils

Lathyrus sylvestris (Everlasting pea, Perennial pea)
Medium seeds: 56 per ounce
Hardy herbaceous perennial
Height up to 6 ft.
Lovely racemes can be as many as 10 flowers which are rose/pink and marked with green and purple
The oval, midgreen leaves have a pair of leaflets attached

Soak seeds in water for 24 hours before sowing or pregerminate on paper towels (see recipe). Sow seeds from autumn until early spring in pots using a standard, soil-less seed mix. Cover with perlite or vermiculite, place under protection at 50–60°F. Germination takes 1–2 weeks. Plant out as soon as all threat of frost has passed, after a period of hardening off.

OR
Sow in spring, any time after the hard frosts 41°F have gone, in prepared drills 2 in. deep. Germination takes 7–14 days. Late spring sowings (May onwards) will produce plants with little flower.

> ### Recipe for sowing, planting, and growing Lathyrus odoratus
> The scent of sweet peas is very evocative; personally they remind me of both my childhood and of flower shows.
>
> #### Ingredients
> 3 seeds per pot
> Sandpaper
> Small bowl or paper towels
> Warm water
> 1 x 12 in. pot
> Standard, soil-less seed mix
> Fine-grade perlite (wetted) or vermiculite
> White plastic plant label
>
> **Method** Whether you sow in pots or directly into the soil, you will get a better rate of germination if you soak the seeds in water before sowing. Before soaking the seeds, scarify them by rubbing gently on sandpaper. This will allow the water to penetrate more easily. Then, either soak the seeds overnight in warm water before sowing, or soak 2 or 3 layers of paper towels on a plate, space the seeds evenly on the paper and cover with 2 or 3 further layers of pre-soaked paper towels. Keep moist, and check daily by lifting the corner of the towel. When the seeds begin to sprout they can be transplanted gently to pots of soil-less mix.
>
> Between autumn and early spring, fill the pots with soil-less mix, smooth over, tap down and water in well. Plant 3 pre-soaked and scarified seeds per pot, spaced equally on the surface of the mix. Gently press the seeds into the mix. Cover the seeds with fine-grade perlite (wetted) or vermiculite and label with the plant name and date. Place the pot in a cool, light place out of direct sunlight, at an optimum temperature of 50–60°F. Germination of seeds that have not sprouted before planting will take 7–14 days. Place the pots in a cold frame or unheated greenhouse. It is important to grow sweet peas in as cool an environment as possible, giving protection only if the weather is very cold. If grown in warm conditions, the young plant will become soft and leggy. It is also necessary to pinch out the growing tips of seedlings sown in winter or spring, although autumn-sown plants should not need it.
>
> Plant out the seedlings in spring, when the temperature does not fall below 40°F, after a period of hardening off, in a site that has been prepared the previous autumn with well-rotted manure. The plants will cover the supports given to them rapidly, so do not skimp on the height, allow a minimum of 6 ft.
>
> Alternatively, you can sow the pre-soaked and scarified seeds directly in a prepared site (see above) in the garden when the nighttime temperature does not fall below 40°F. The site must be prepared well in early autumn. Space the seeds 18 in. apart, next to some support, either canes or a frame, minimum height 6 ft. Press gently into the soil and cover lightly, water in well. Germination takes 7–14 days.

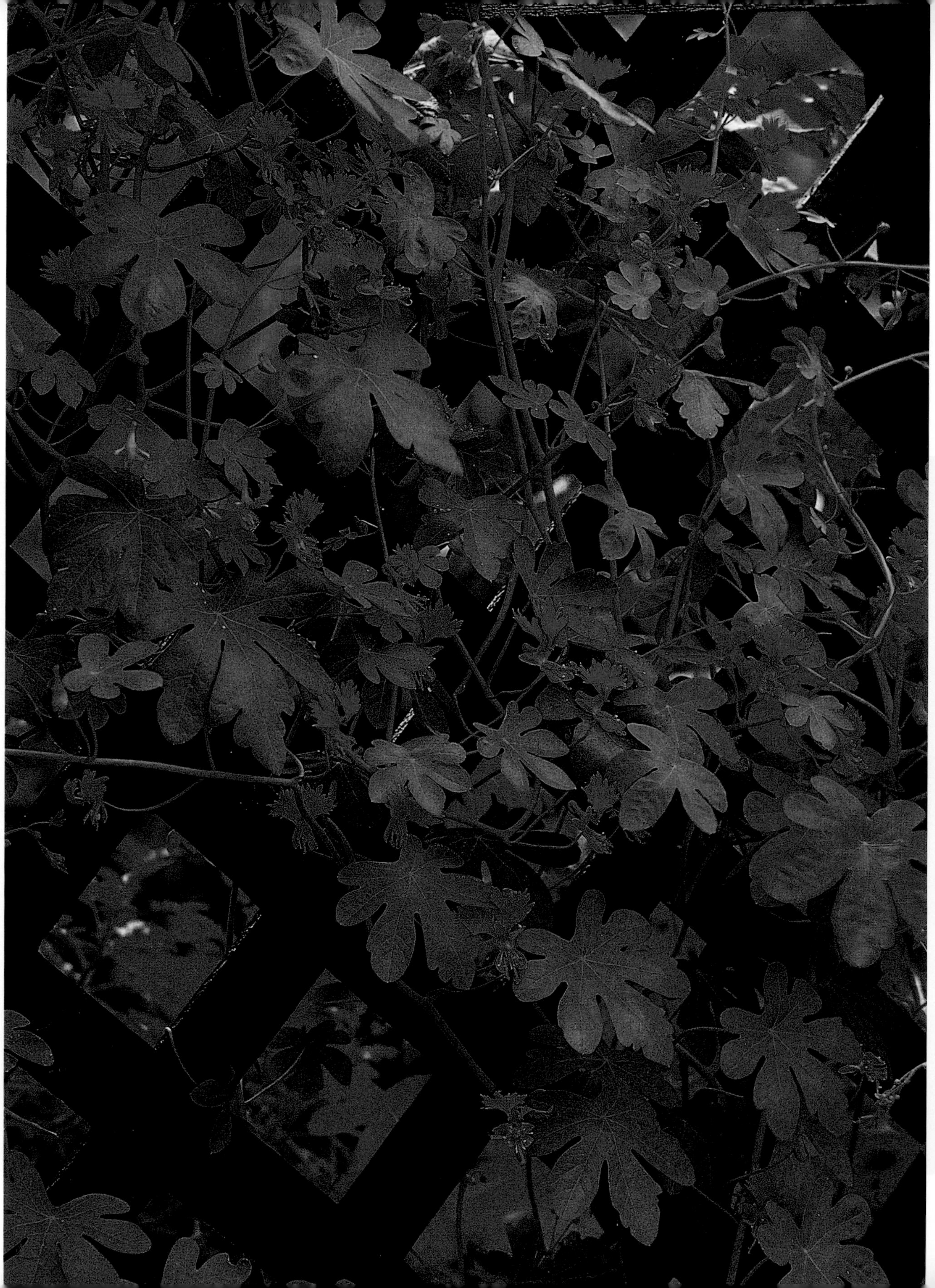

Tropaeolum peregrinum

Rhodochiton atrosanguineum (Purple bell vine, Rhodochiton)
Small seeds: 9,800 per ounce
Evergreen, tender perennial, grown as an annual
Height up to 10 ft.
Tubular, dark maroon/purple flowers, surrounded by bell-shaped, red/purple calyces
Flowers all summer
Oval, toothed, midgreen leaves

Sow fresh seeds in early spring in pots or cell packs, using a standard, soil-less seed mix. Cover with perlite or vermiculite, and place under protection at 70°F. Germination takes 6–10 days.

Solanum *Solanaceae*

A large genus of annuals, perennials, evergreens, shrubs, and climbers. Plant in a rich, moist, neutral soil in sun or partial shade.

Solanum dulcamara (Woody nightshade—not Deadly nightshade)
POISONOUS PLANT Zones 5–9
Medium seeds: 280 per ounce
Hardy perennial
Height up to 6 ft
Small, exotic looking, purple and yellow flowers, followed by green berries that turn scarlet in winter
Flowers in summer
Large, green, oval, pointed leaves

Sow seeds ,extracted from berries, in autumn in pots or cell packs, using standard loam-based seed mix, mixed with coarse horticultural sand. Mix in a ratio of 1 part soil mix + 1 part sand. Cover with coarse horticultural sand, then place outside exposed to all weathers (see "Breaking Seed Dormancy", page 233, for more information). Germination takes 4–6 months but can be erratic, so be patient and do not discard container.

Thunbergia *Acanthaceae*

A genus of half-hardy annual and perennial evergreen climbers. Plant in any fertile, well-drained soil in sun or partial shade.

Thunbergia alata (Black-eyed Susan vine)
Medium seeds: 112 per ounce
Tender annual
Height up to 10 ft
Pretty, orange-yellow flowers with very dark-brown centers
Heart-shaped, midgreen leaves with toothed edges

Sow fresh seeds in early spring in pots or cell packs, using a standard, soil-less seed mix. Cover with perlite or vermiculite and place under protection at 70°F. Germination takes 7–14 days.

Tropaeolum *Tropaeolaceae*

A genus of hardy to frost-tender annuals, perennials, and climbers; all have lovely flowers. Plant in well-drained soil in a sunny position.

Tropaeolum peregrinum (Canary creeper)
Large seeds: 22 per ounce
Tender perennial, often grown as an annual
Height up to 6 ft.
Single, small, bright yellow flowers in summer until the first frosts
Gray green, lobed leaves

Sow fresh seeds in early spring in pots or cell packs, using a standard, soil-less seed mix. Cover with perlite or vermiculite, place under protection at 65°F. Germination takes 5–12 days.

Tropaeolum speciosum (Flame creeper, Flame nasturtium)
Zones 8–9
Medium seeds: 56 per ounce
Hardy perennial
Height up to 10 ft
Scarlet flowers, followed by bright blue fruits
Flowers in summer
Blue-green lobed leaves
The roots of this plant like to be kept in the shade so that they do not dry out in hot summers

Sow fresh seed in autumn in pots or cell packs, using standard, loam-based seed mix. Cover lightly with soil mix and place in a cold frame. Germination takes 3–4 weeks. If germination does not occur during this time, place container outside exposed to all weathers (see "Breaking Seed Dormancy", page 233) for more information). Germination takes 5–7 months on average, but can take as long as 2 years. When germination has occurred, overwinter young plants in a cold frame. If you are using old seeds, soak them for 24 hours before sowing. Flowers after 3 years or, occasionally, after as long as 5 years.

Wisteria *Papilionaceae*

A genus of hardy, deciduous climbers, grown for their lovely flowers. Plant in fertile, well-drained soil in a sunny position.

Wisteria sinensis (Wisteria)
Large seeds: 8 per ounce
Hardy deciduous perennial
Height up to 40 ft.
Sweetly scented, mauve/purple, pea-like flowers, which hang in clusters
Flowers in early summer
Divided, midgreen leaves with oval leaflets

Soak seeds for 24 hours before sowing. Sow seeds in spring in pots or cell packs, using standard, soil-less seed mix, either peat or a peat substitute. Cover with perlite or vermiculite. Place in a cold frame. Germination takes 1–6 months. Wisteria grown from seed will take up to 7 years before flowering. Plants raised in this way are often used as root stock for grafting.

ferns

Ferns are beautiful plants that come in a wide variety of forms and textures. They are excellent for the shady garden. Ferns are primitive plants that do not flower but reproduce sexually from spores, rather than seeds. There are two distinct stages in the life cycle of a fern. During the first stage, the mature plants produce spores on the underside of their leaves. During the second stage, the spores are dispersed from the fern and they germinate. After germination they grow into small, heart-shaped plants known as prothalli. Male and female cells are produced on these plants and, after fertilisation, the adult fern begins to develop.

Dicksonia antartica shoots, see page 89

Collecting spores, sowing method, and potting on

To collect spores

In mid- to late summer, place a portion of mature fern frond on a piece of paper in a dry place. If spores are ripe, they will be shed onto the paper and will appear as black, brown or yellow "powder", which is a mixture of spores and fragments of the spore cases (sporangia).

Sowing method

It is important to sterilize the germination mixture before sowing the fern spores. Do this by pouring boiling water over it. This kills the spores of fungi and other plants that may germinate and crowd out the developing fern prothalli.

Spores should be sprinkled sparsely on a suitable medium, such as finely chopped tree-fern fiber, peat moss or sphagnum moss. Two parts peat moss mixed with one part coarse grit also forms an excellent germination base.

Once sown, the containers should be covered with plastic wrap, plastic or glass (allowing some air space) and kept in indirect light at around 68°F for cool-temperate ferns and 80°F for tropical ferns. Spores take from 2 to 6 weeks to germinate.

Potting on

After a few weeks, the germinating spores appear as a velvety green haze on the surface of the sowing mix. If it is slimy, there may be algae contamination. In this case I recommend that you discard it, because contaminated spores are difficult to rescue. Next time, make sure everything is spotlessly clean, taking the extra precaution of sterilizing the sowing mix with boiling water before sowing.

The spring following sowing, when the prothalli are formed and well-developed, fill some spotlessly clean pots with the sterile, soil-less seed mix. With a clean knife, lift small pieces of prothalli and place them gently on the surface of the soil-less mix. Spray with sterilized water, cover with plastic wrap, plastic or glass and place back under protection. The container should be covered with glass or plastic until the fern fronds appear. When the young fronds are growing well, harden off gradually by admitting more light and air. When they are 3 in. tall, divide and pot up individually, using a mix of two parts peat moss with one part coarse grit, then grow on in partial shade. The hardy varieties of fern will be large enough to plant out in 2–3 years.

Matteuccia *Woodsiaceae*

A genus of hardy, rhizomatous, deciduous ferns. Plant in wet soil in semishade.

Matteuccia struthiopteris (Ostrich fern) Zones 2–7
Propagated by spores
Hardy deciduous fern
Height 3 1/2 ft.
Lance-shaped green divided fronds that look rather like ostrich feathers and grow in a shuttlecock habit

Sow fresh spores at 60°F following the instructions on the left.

Platycerium *Polypodiaceae*

A genus of tender, evergreen ferns. These are best grown in hanging baskets.

Platycerium bifurcatum AGM (Stag's-horn fern)
Propagated by spores
Tender evergreen fern
Minimum temperature 40°F
Height 3 1/2 ft
Broad sterile fronds and long, arching, forked, gray-green fertile fronds

Sow fresh spores at 70°F following the instructions on the left.

Polypodium *Polypodiaceae*

A genus of tender to hardy semievergreen or evergreen ferns. Plant in moist but not wet soil in semishade.

Polypodium vulgare (Common polypody) Zones 5–9
Propagated by spores
Hardy evergreen fern
Height 12 in.
Narrow, lance-shaped, divided, midgreen fronds

Sow fresh spores at 60°F following the instructions on the left.

Thelypteris *Thelypteridaceae*

A genus of hardy, deciduous ferns. Plant in moist soil in semishade.

Thelypteris palustris (Marsh fern) Zones 3–8
Propagated by spores
Hardy deciduous fern
Height 2 1/2 ft.
Long, lance-shaped, pale green fronds, which are very divided

Sow fresh spores at 60°F following the instructions on the left.

grasses

There is a huge range of plants in the grass family, from the common lawn grass to fashionable bamboos and ornamental grasses. They are excellent in the garden, giving color, texture, sound, and graceful movement. Another benefit of growing ornamental grasses is that they dry wonderfully for winter flower arrangements.

Grasses are very rewarding plants to grow from seed, and the seeds are easy to collect. In the late summer and early autumn, pick the grass stem with the flower, allowing the bracts, which hold the seed, to fluff up or dry out fully. This will take a few days. Strip the seeds from each stem. As most grasses do not need a period of stratification, either sow immediately or store until spring.

Cortaderia selloana, see page 96

Bouteloua *Poaceae*

A genus of annuals, perennials, and semievergreens. Plant in any well-drained soil in sun to partial shade.

Bouteloua gracilis (Blue grama, Mosquito grass) Zones 3–9
Medium seeds: 2100 per ounce
Hardy perennial, semievergreen
Height 15 in.
Comb- or mosquito-shaped flowering spikes that hang at right angles to the stem
Flowers in summer
Decorative, narrow, nodding, brown green leaves that grow in tufts
Good for flower arrangements, fresh or dried

Sow in late summer/early autumn in prepared open ground when air temperature does not go below 50°F at night. Germination takes 14–21 days.
OR
Sow in spring in pots or cell packs, using standard, soil-less seed mix, either peat or a peat substitute. Place under protection at 59°F. Cover with perlite or vermiculite. Germination takes 7–14 days.

Briza *Poaceae*

A genus of annuals, perennials, and semievergreens. Plant in any well-drained soil in sun to partial shade. This genera self-seeds freely.

Briza maxima (Greater quaking grass)
Medium seeds: 1,120 per ounce
Hardy annual
Height up to 20 in.
Loose panicles of pendant, oval-shaped, purplish-green spikes
Panicles borne in summer, up to 10 panicles on a single stem
Midgreen, narrow leaves

Briza media (Common quaking grass) Zones 4–9
Small seeds: 9,240 per ounce
Hardy perennial
Height up to 2 ft.
Open panicles of pendant, heart-shaped, purplish-brown spikes
Panicles borne in summer, up to 30 panicles on a stem
Midgreen narrow leaves at the base of the plant
Lovely in dry flower arrangements

Sow in late summer to early autumn in prepared open ground, when air temperature does not go below 48°F at night. Germination takes 10–15 days.
OR
Sow in spring in pots or cell packs, using standard, soil-less seed mix. Place under protection at 50°F, cover with perlite or vermiculite. Germination takes 7–14 days. Will flower within 14 weeks of sowing.

Bromus *Poaceae*

A genus of mainly annuals and some perennials. Plant in any well-drained soil in sun to partial shade.

Bromus macrostachys
Medium seeds: 700 per ounce
Hardy annual
Height up to 24 in.
Large spikes with spreading awns
Awns borne in early summer
Narrow compact leaves

Sow in late summer to early autumn in prepared open ground, when air temperature does not go below 48°F at night. Germination takes 10–15 days.
OR
Sow in spring in pots or cell packs, using standard, soil-less seed mix. Place under protection at 50°F, cover with perlite or vermiculite. Germination takes 7–14 days.

Carex *Cyperaceae*

A genus of evergreen, rhizomatous perennials from the sedge family. Grows naturally by water, but will adapt to any soil, in sun or partial shade.

Carex buchananii (Leatherleaf sedge) Zones 7–9
Medium seeds: 1,960 per ounce
Hardy evergreen
Height up to 2 ft.
Rather insignificant, brown, spikelet flowers
Flowers in summer
Very narrow, copper-coloured leaves, which turn red towards the base, making an attractive mound

Carex pendula (Pendulous sedge) Zones 8–10
Medium seeds: 1,960 per ounce
Hardy evergreen
Height up to 3$\frac{1}{2}$ ft.
Pendant green/brown flower spikes, which look rather like a drooping cat's tail
Flowers in summer
Narrow, midgreen leaves

Sow in late summer/early autumn in prepared open ground, when air temperature does not fall below 48°F at night. Germination takes 10–15 days.
OR
Sow in spring in pots or cell packs, using standard, soil-less seed mix. Place under protection at 50°F, cover with perlite or vermiculite. Germination takes 7–14 days.

Chasmanthium *Poaceae*

A genus of annuals, perennials, and semievergreens. Plant in any well-drained soil in sun to partial shade.

Chasmanthium latifolium (*Uniola latifolia*) (Wood oats) Zones 5–9
Medium seeds: 336 per ounce
Hardy perennial
Height 2¹/₂ ft.
Attractive drooping panicles of flat brown heads
Panicles borne in summer the first year after sowing
Tufted, midgreen flat leaves

Sow in late summer/early autumn in prepared, open ground, when air temperature does not fall below 50°F at night. Germination takes 14–21 days.
OR
Sow in spring in pots or cell packs, using standard, soil-less seed mix. Place under protection at 59°F, cover with perlite or vermiculite. Germination takes 7–14 days.

Cordyline *Agavaceae*

A genus of half-hardy shrubs and trees. Plant in fertile well-drained soil in sun or partial shade. Makes an impressive potted plant, cut watering down in the winter.

Cordyline australis (New Zealand cabbage palm) Zones 8–10
Medium seeds: 700 per ounce
Half-hardy perennial
Height 10 ft. In the correct conditions it will grow up to 33 ft.
Fragrant, creamy-white flowers, followed by small black berries
Mature plants flower in summer
Dense clusters of narrow, sword-like, gray green leaves
Makes a great feature as a large potted plant

Cordyline indivisa (Cabbage Palm, Dracaena) Zones 9–10
Medium seeds: 560 per ounce
Half-hardy perennial, often grown as an annual
Height 10 ft.
Clusters of tiny, star-shaped, white flowers
Flowers in summer on mature plants only
Long, 2–6 ft., tapering, midgreen leaves with a midrib of red or yellow
Makes a good feature as a potted plant

In their countries of origin, both the above are small trees. Away from their native conditions, they make lovely container plants, which displays their distinctive leaves to great advantage.

Sow in spring in pots or cell packs, using standard, soil-less seed mix. Place under protection at 70°F, cover with perlite or vermiculite. Germination takes 30–40 days.

Recipe for sowing, planting, and growing **Cordyline**
Cordylines are not grasses or palms. As their common name suggests, they are evergreen trees or shrubs. They are grown in many countries, however, as a foliage plant and, for this reason, they have been grouped with the grasses. They look tremendous planted out during the growing season, their leaves give a texture and structure to the garden. In containers they can form a lovely, striking center piece.

Ingredients
4 seeds per cell or 6 seeds per pot
1 flat with cell packs
OR
1 x 4 in. pot
Standard soil-less seed mix
Fine-grade perlite (wetted) or vermiculite
White plastic plant label

Method Fill the flat or pot with soil-less mix, smooth over, tap down and water in well. Sow the seeds thinly on the top of the compost and press in gently with the palm of the hand. Cover with perlite or vermiculite, and label with the plant name and date. Place the flat or pot in a warm light place, out of direct sunlight, at an optimum temperature of 70°F. Keep watering to a minimum until germination has taken place, which takes 30–40 days in spring with warmth. Shortly after germination has taken place, and when the seedlings have emerged fully, put the containers in a cooler environment at 65°F. You can prick out and pot up approximately 4–6 weeks after germination. These young plants will need a full season's growth with winter protection at not less than 50°F before planting out as a summer bedding display. If you are growing the plants on in containers, you will need to split the seedlings as soon as they are large enough to handle. Replant in 4 in. pots, using a soil-less potting mix.

Cortaderia *Poaceae*

A genus of annuals, perennials, and semievergreens. Plant in any well-drained soil in sun to partial shade.

Cortaderia selloana (Pampas grass) Zones 8–10
Small seeds: 14,000 per ounce
Hardy evergreen perennial
Height up to 8 ft.
Very attractive plumes of silver panicles
Panicles borne in summer
Narrow, sharp-edged, midgreen leaves, which can grow to 5 ft. long

Sow in spring in pots or cell packs, using standard, soil-less seed mix. Place under protection at 65°F, cover with perlite or vermiculite. Germination takes 14–21 days.

Eragrostis *Poaceae*

A genus of annuals, perennials, and semievergreens. Plant in any well-drained soil in sun to partial shade.

Eragrostis curvula (Weeping love grass) Zones 7–9
Small seeds: 3,360 per ounce
Hardy annual
Height up to 4 ft.
Drooping panicles of dark olive gray flowers
Flowers all summer
Impressive mounds of fine, dark-green, arching leaves

Sow in spring in pots or cell packs, using standard soil-less seed mix. Cover with perlite or vermiculite and place under protection at 70°F. Germination takes 4–10 days.

Festuca *Poaceae*

A genus of annuals, perennials, and semievergreens. Plant in any well-drained, poor soil in a sunny situation.

Festuca glauca (Blue fescue) Zones 4–9
Small seeds: 3,640 per ounce
Hardy evergreen perennial
Height up to 10 in.
The flowers are inconspicuous spikelets
Flowers in summer
Narrow leaves in shades of blue green
This plant is very good for edging

Sow in late summer/early autumn in prepared open ground, when air temperature does not go below 50°F at night. Germination takes 14–21 days.
OR
Sow in spring in pots or cell packs, using standard, soil-less seed mix and cover with perlite or vermiculite. Place under protection at 59°F. Germination takes 7–14 days.

Festuca glauca

Hordeum *Poaceae*

A genus of annuals, perennials, and semievergreens. Plant in any well-drained soil in sun to partial shade.

Hordeum jubatum (Squirrel's-tail grass) Zones 4–9
Medium seeds: 1,540 per ounce
Perennial often grown as an annual
Height up to 2 ft.
Lovely feathery, plume-like, arching flower spikes with a silky beard at the end of the grain sheath
Flowers in summer
Branching stems with narrow, dark-green leaves
This plant looks lovely in gentle winds

Sow in late summer/early autumn in prepared open ground, when air temperature does not go below 50°F at night. Germination takes 14–21 days.
OR
Sow in spring in pots or cell packs, using standard soil-less seed mix. Place under protection at 59°F, cover with perlite or vermiculite. Germination takes 7–14 days.

Koeleria *Poaceae*

A genus of annuals, perennials, and semievergreens. Plant in any well-drained soil in sun to partial shade.

Koeleria glauca (Glaucous hair grass) Zones 6–9
Small seeds: 7,000 per ounce
Hardy perennial
Height up to 12 in.
Attractive, upright, gray blue flower heads
Flowers in summer
Dense tussocks of narrow, blue gray leaves

Sow in late summer/early autumn in prepared open ground, when air temperature does not go below 48°F at night. Germination takes 10–15 days.
OR
Sow in spring in pots or cell packs, using standard soil-less seed mix. Place under protection at 70°F, cover with perlite or vermiculite. Germination takes 3–7 days.

Lagurus *Poaceae*

A genus of annuals, perennials, and semievergreens. Plant on any well-drained site in a sunny situation, although it prefers a sandy soil.

Lagurus ovatus (Hare's-tail grass)
Small seeds: 6,720 per ounce
Hardy annual
Height 18 in.
Very attractive, egg-shaped, soft panicles of white flower spikes with gold stamens
Flowers in summer
Flat, long, narrow, midgreen leaves

Sow in late summer/early autumn in prepared open ground, when air temperature does not go below 50°F at night. Germination takes 14–21 days.
OR
Sow in spring in pots or cell packs, using standard soil-less seed mix. Place under protection at 68°F, cover with perlite or vermiculite. Germination takes 7–14 days.

***Recipe for sowing, planting, and growing* Lagurus ovatus**
This attractive annual grass looks lovely in the garden grown in clumps or drifts. It is equally happy grown as a container plant.

Ingredients
6 seeds per cell or 18 seeds per pot
1 white 3 x 5 card, folded in half
1 flat with cell packs
OR
1 x 4 in. pot
Standard soil-less seed mix
Fine-grade perlite (wetted) or vermiculite
White plastic plant label

Method Fill the cell packs or pot with soil-less mix, smooth over, tap down and water in well. As the seeds are very fine, use the folded card to help you to sow them thinly (see page 240). Cover with perlite or vermiculite and label with the plant name and date. Place the flat or pot in a warm, light place, out of direct sunlight, at an optimum temperature of 68°F. Keep watering to a minimum until germination has taken place, which takes 7–10 days with warmth. Shortly after germination, when the seedlings have emerged fully, put the containers in a cooler environment at 58°F. Spring-sown plants can be pricked out and potted up approximately 4 weeks after germination. The spring-sown seedlings will be ready to plant out approximately 4 weeks after that. If you are using cell packs, you can plant out directly into a container or the garden as soon as they have rooted right down the cell pack. If you are growing this grass as a container plant, you will need to pot it up in a5 in. diameter pot. Otherwise, you can divide the seedlings up and repot using a soil-less potting mix. Keep the container protected until all threat of frost has gone, then harden off before leaving out all night. If you wish to plant in the garden, make sure the seedlings have a period of hardening off, then plant out when there is no threat of frost. Plants grown with heat can bloom within 16 weeks of sowing.

Alternatively, you can sow directly in a prepared site in the garden when the nighttime temperature does not fall below 50°F. Sow the fine seeds thinly, either in a short row, or in a round clump, press gently into the soil with the palm of the hand, and cover lightly. Water in well, using a fine rose on the watering can so that you do not disturb the fine seeds. Germination takes 14–21 days. If you have sown too thickly, it is a good idea to thin the seedlings. Alternatively, you can wait until the seedlings have established, then divide them and replant in the garden.

Luzula *Juncaceae*

A genus of hardy annuals, perennials, and evergreens from the rush family. Plant in a moist but free-draining soil in sun or partial shade.

Luzula nivea (Snowy woodrush) Zones 6–9
Small seeds: 4,480 per ounce
Hardy evergreen perennial
Height up to 2 ft.
Dense clusters of white flower spikes
Flowers in early summer
Midgreen leaves with white, hairy edges

Luzula sylvatica (Greater woodrush) Zones 4–9
Small seeds: 7,560 per ounce
Hardy evergreen perennial
Height up to 12 in
Brown flower spikes
Flowers in summer
Midgreen leaves with hairy edges that grow in thick tufts

Sow in spring in pots or cell packs, using standard soil-less seed mix. Place under protection at 70°F, cover with perlite or vermiculite. Germination takes 3–7 days.
   These two decorative rushes are very useful in the garden, as both like being planted in partial shade.

Pennisetum *Poaceae*

A genus of annuals, perennials, and semievergreens. Plant in any well-drained soil in sun to partial shade.

Pennisetum setaceum (Fountain grass)
Small seeds: 4,480 per ounce
Half-hardy perennial grown as an annual
Height up to 3¹/₂ ft.
Dense cylindrical panicles of copper red spikes with attractive bearded bristles
Panicles borne from midsummer until early winter
Tall, fine, tufted midgreen leaves

Pennisetum villosum (Feather-top)
Medium seeds: 1,400 per ounce
Half-hardy perennial grown as an annual
Height up to 3¹/₂ ft.
Panicles of soft, creamy-pink spikes, which fade to pale brown, with attractive, long-bearded bristles
Panicles borne from midsummer until early frosts
Tufted, midgreen leaves

Both of these species need winter protection in cold, wet climates. Place in a cold frame or cold greenhouse at a minimum temperature of 40°F at night.

Sow in late summer/early autumn in prepared open ground, when air temperature does not go below 48°F at night. Germination takes 10–15 days.

OR
Sow in spring in pots or modules, using standard soil-less seed mix. Place under protection at 70°F, cover with perlite or vermiculite. Germination takes 4–8 days.

Phalaris *Poaceae*

A genus of annuals, perennials, and semievergreens. Plant in any well-drained soil in sun to partial shade.

Phalaris canariensis (Canary grass)
Medium seeds: 364 per ounce
Hardy annual
Height 1–4 ft.
The flower heads are white with green veins that are often tinged bright violet
Flowers in summer
Broad, midgreen leaves
Canary grass was introduced from the Canary Islands to feed canaries

Sow in late summer/early autumn in a prepared open ground, when air temperature does not go below 50°F at night. Germination takes 14–21 days.
OR
Sow in spring in pots or cell packs, using standard soil-less seed mix. Place under protection at 59°F. Cover with perlite or vermiculite. Germination takes 7–14 days.

Polypogon *Poaceae*

A genus of annuals, perennials, and semievergreens. Plant in a moist/damp soil in sun to partial shade.

Polypogon monspeliensis (Rabbit's-foot grass, Annual beard grass)

Small seeds: 19,600 per ounce
Hardy annual
Height up to 2 ft.
Dense tawny panicles that resemble a rabbit's foot, soft and bristly
Panicles borne in summer
Flat, midgreen leaf, fine and rough to the touch
Grows happily at water margins and along ditches

Sow in late summer/early autumn in a prepared open ground, when air temperature does not go below 48°F at night. Germination takes 10–15 days.
OR
Sow in spring in pots or cell packs, using standard soil-less seed mix. Place under protection at 50°F. Cover with perlite or vermiculite. Germination takes 7–14 days.

Setaria *Poaceae*

A genus of annuals, perennials, and semievergreens. Plant in any well-drained soil in sun to partial shade.

Setaria macrochaeta (Fox-tail grass)
Medium seeds: 1,400 per ounce
Hardy annual
Height up to 3 ft.
Long flower panicles adorned with long, pendulous bristles
Panicles borne in summer
Lance-shaped, midgreen leaves

Sow in late summer/early autumn in prepared open ground, when air temperature does not go below 50°F at night. Germination takes 14–21 days.
OR
Sow in spring in pots or cell packs, using standard soil-less seed mix. Place under protection at 59°F. Cover with perlite or vermiculite. Germination takes 7–14 days.

Stipa *Poaceae*

A genus of annuals, perennials, and semievergreens. Plant in any well-drained soil in sun to partial shade.

Stipa gigantea (Giant feather grass) Zones 6–9
Medium seeds: 98 per ounce
Hardy evergreen perennial
Height up to 8 ft.
A very elegant grass with open panicles of golden spikes, long awns and dangling golden anthers
Panicles borne from summer until late autumn
Narrow midgreen tufted grass

Stipa tenuissima (Hair grass) Zones 6–9
Medium seeds: 2,100 per ounce
Hardy evergreen perennial
Height up to 20 in.
Arching flower heads, which start greenish-white then change to a buff color
Flowers in summer
Erect, light clumps of thin pale yellow-green leaves

Sow (where hardy) in late summer/early autumn in prepared open ground, when air temperature does not go below 50°F at night. Germination takes 14–21 days.
OR
Sow in spring in pots or cellpacks using standard soil-less seed mix. Place under protection at 59°F. Cover with perlite or vermiculite. Germination takes 7–14 days.

Thamnocalamus *Bambusoideae*

A genus of hardy to half-hardy perennials, some of which are evergreens. Plant in a well-drained soil, which can be poor. A good mulch of organic matter in spring will feed the plant and retain some moisture. Prefers a sheltered, sunny or semishaded position.

Thamnocalamus spathaceus (*Fargesia murielae*) (Muriel bamboo, Umbrella bamboo) Zones 4-9
Medium seeds: 280 per ounce
Hardy evergreen perennial
Height up to 12 ft.
Rather boring flower spikes
Flowers in summer
Attractive young stems with loose, light brown sheaths
Broad, lance-shaped, apple-green leaves

Sow in spring or early autumn in pots or cell packs, using standard, soil-less seed mix. Place under protection at 41°F. Cover with perlite or vermiculite. Germination can be very irregular and periodic over a long period, so do not give up. Do not allow the soil-less mix to dry out.

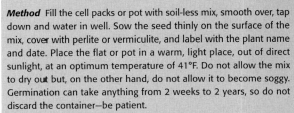

***Recipe for sowing, planting, and growing* Thamnocalamus spathaceus**
Bamboos are very attractive and graceful plants, which look lovely growing in beds. This species is no exception, with its broad, apple-green leaves.
Ingredients
3 seeds per cell or 5 seeds per pot
1 flat with cell packs
OR
1 x 4 in. pot
Standard soil-less seed mix
Fine-grade perlite (wetted) or vermiculite
White plastic plant label

Method Fill the cell packs or pot with soil-less mix, smooth over, tap down and water in well. Sow the seed thinly on the surface of the mix, cover with perlite or vermiculite, and label with the plant name and date. Place the flat or pot in a warm, light place, out of direct sunlight, at an optimum temperature of 41°F. Do not allow the mix to dry out but, on the other hand, do not allow it to become soggy. Germination can take anything from 2 weeks to 2 years, so do not discard the container—be patient.

Once the seeds have germinated and are large enough to handle, either pot up from cell packs into pots or, if seeds have been sown in a pot, divide the seedlings and repot in pots using a soil-based potting mix. Keep the container protected until all threat of frost has gone, then harden off before leaving out all night. If you wish to plant in the garden, make sure the seedlings have a period of hardening off, then plant out when there is no further threat of frost.

Herbs are some of the most rewarding plants you can grow in any garden, they look good, taste good and do you good. They will adapt to most growing conditions and many are happy growing in containers. What more can you ask from a plant? Most culinary herbs originate from either the Mediterranean or the Tropics, so, for those living in cool or cold climates, do not be in too much of a hurry to sow the seed in spring, wait until the weather has warmed up and the light levels increase.

herbs

Myrrhis odorata, see page 121

Agrimonia *Rosaceae*

A genus of herbaceous, rhizomatous perennials with long spikes of yellow flowers. Mainly medicinal herbs and also dye plants. Plant in a sunny situation in a well-drained alkaline soil.

Agrimonia eupatoria (Agrimony, Sticklewort) Zones 6–9
Medicinal herb
Medium seeds: 322 per ounce
Hardy perennial
Height 1–2 ft.
Small, yellow, lightly scented flowers
Flowers in summer
Elliptic to oval, toothed and hairy green leaves

Sow fresh seed in autumn in pots or cell packs, using standard loam-based soil mix. Cover lightly with soil mix, place in a cold frame. Germination takes 3–4 weeks. If no germination occurs during this time, place the container outside, exposed to all weathers (see "Breaking Seed Dormancy", page 233, for more information). When germination has taken place, overwinter young plants in the cold frame. Can flower in the first year.

Allium *Alliaceae*

This large genus of perennials, some of which are edible, can come in different forms, bulbs, rhizomes o,r fibrous rootstock, but nearly all have onion-smelling leaves, and most have small flowers, which are clustered together in spherical or similar shapes. They are all hardy and, in the majority of cases, require a sunny site with a well-drained, rich soil.

Allium fistulosum (Welsh onion, Japanese leek) Zones 5–9
Culinary and medicinal herb
Medium seeds: 14,000 per ounce
Hardy perennial
Height 20 in.
Large, creamy-white global flowers
Flowers in spring to summer of the second year
Long, green, hollow cylindrical leaves

Allium schoenoprasum (Chives) Zones 4–9
Culinary herb
Medium seeds: 2,240 per ounce
Hardy perennial
Height 12 in.
Purple globular flowers
Flowers all summer
Leaves green, hollow, and cylindrical

Allium tuberosum (Garlic chives) Zones 5–9
Culinary herb
Medium seeds: 560per ounce
Hardy perennial
Height 12 in)
White, star-shaped flowers
Flowers in summer
Flat, solid, thin, lance-shaped, bristly midgreen leaves

Sow in spring in pots or cell packs, using standard soil-less seed mix, either peat or a peat substitute. Cover with perlite or vermiculite, place under protection at 68°F. Germination takes 10–20 days.
OR
Sow in late spring in prepared open ground, when the air temperature does not go below 45°F at night. Germination takes 14–20 days.

Allium ursinum (Ramsons, Wild garlic) Zones 5–9
Culinary and medicinal herb
Medium seeds: 560 per ounce
Perennial
Height 12–18 in.
Clusters of white, star-shaped flowers
Flowers in late spring
Elliptic midgreen leaves

Sow seeds in autumn in pots or cell packs using a standard loam-based seed mix, mixed with coarse horticultural sand. Mix to a ratio of 1 part soil mix + 1 part sand. Cover with coarse horticultural sand and then place outside exposed to all weathers (see "Breaking Seed Dormancy", page 233, for more information). Germination takes 4-6 months, but it can be erratic, so be patient and do not discard the container.
OR
Sow in early autumn in prepared open ground. Mark the row or area clearly. Germination takes place the following spring, but can be erratic.

Anchusa *Boraginaceae*

A genus of annuals, biennials, or perennials, usually with attractive flowers. Plant in a sunny position in a well-drained soil. Dislikes wet winters.

Anchusa officinalis (Alkanet) Zones 4–9
Medicinal herb
Medium seeds: 644 per ounce
Hardy perennial, sometimes grown as a biennial
Height 3¹/₂ ft.
Bright blue flowers, similar to forget-me-nots
Flowers in early summer
Rough lance-shaped evergreen leaves.

Anchusa capensis "Blue Angel"
Medicinal herb
Medium seeds: 1,400 per ounce
Hardy biennial grown as an annual
Height 8 in.
Small clusters of brilliant blue flowers
Flowers in summer
Lance-shaped bristly leaves

Sow fresh perennial seeds in autumn in pots or cell packs, using standard soil-less seed mix, either peat or a peat substitute. Cover with mix, place in a cold frame. Germination takes 2–16 weeks. Overwinter young plants in a cold frame.

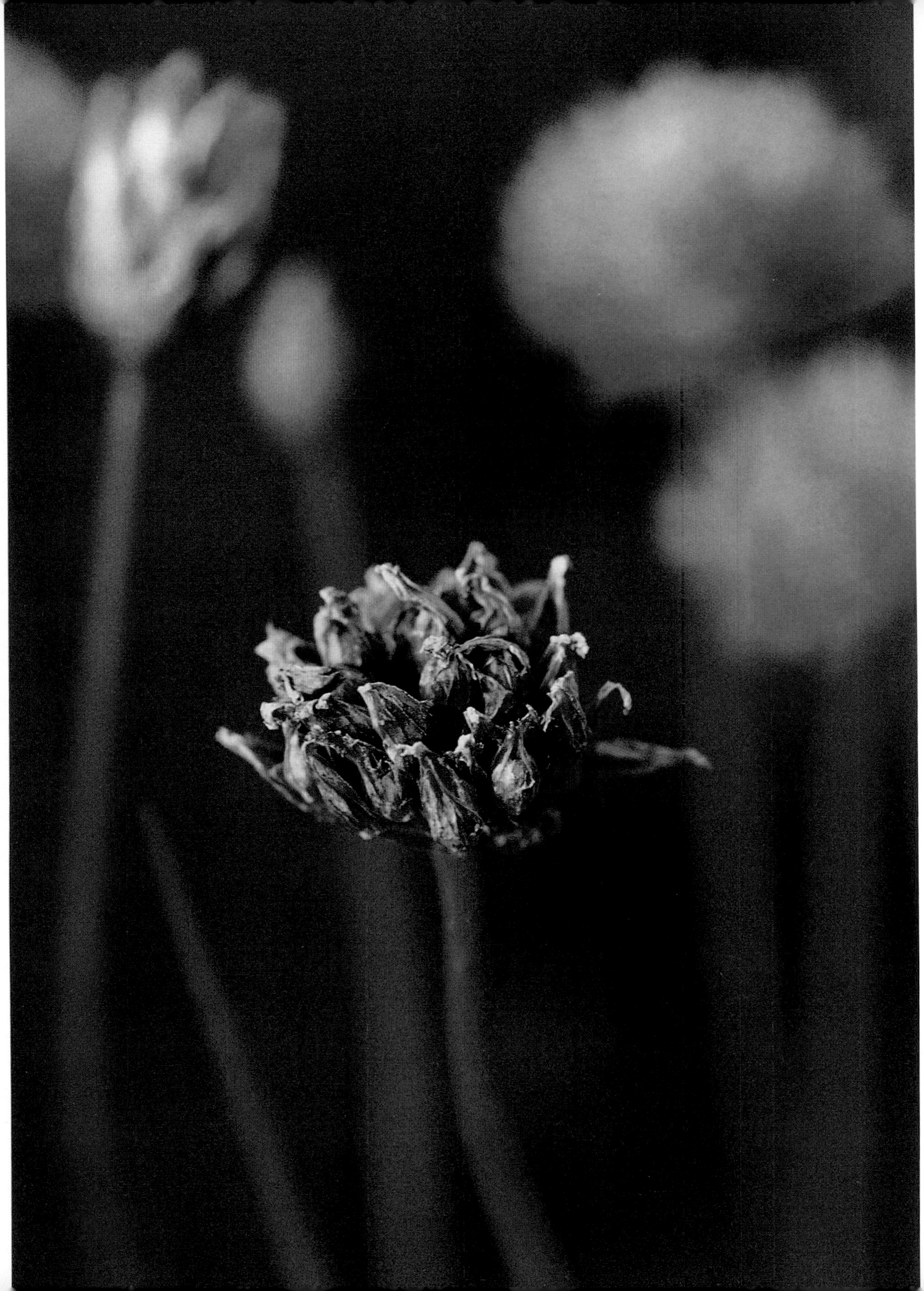

Allium schoenoprasum

OR

Sow biennial seeds in spring in pots or cell packs, using standard soil-less seed mix, either peat or a peat substitute. Cover with mix, place in a cold frame. Germination takes 2-3 weeks.

OR

Sow biennial seeds in late spring in prepared open ground, when the air temperature does not go below 45°F at night. Germination takes 2-3 weeks

Anethum *Apiaceae*

There is only a single species in this genus, mentioned below. Plant in a sunny site in a well-drained soil. Do not plant by fennel as they can hybridize. The seed is viable for 3–5 years.

Anethum graveolens (Dill)
Culinary and medicinal herb
Medium seeds: 1,680 per ounce
Hardy annual
Height 3 ft.
Umbels of small yellow flowers, followed by oval, flat, aromatic seeds
Flowers in summer
Fine, thread-like, midgreen aromatic leaves

Sow seeds in early spring in pots, using standard soil-less seed mix, either peat or a peat substitute. Cover with perlite or vermiculite, place under protection at 60°F. Germination takes 5–10 days.

OR

Sow seeds in late spring in prepared open ground, when the air temperature does not go below 45°F at night. Germination takes 2–3 weeks.

In my opinion, sowing direct in open ground gives the best crop, because dill hates being transplanted.

Angelica *Apiaceae*

A genus of short-lived perennials and biennials, which have many umbels of small, white to purple flowers in summer. The ideal planting situation would be with the roots in the shade and the flowers in the sun. Angelica prefers a damp soil.

Angelica archangelica (Angelica) Zones 4–9
Culinary and medicinal herb
Medium seeds: 560 per ounce
Biennial
Height 6 ft.
Umbels of small, green-white flowers
Flowers in late spring of the second year
Large, deeply divided, midgreen leaves

It is vital to sow fresh seeds, as seeds are viable for only 3 months. Sow seed in autumn in pots or cell packs, using a standard loam-based seed mix, mixed with coarse horticultural sand to a ratio of 1 part soil mix + 1 part sand. Cover with coarse horticultural sand, then place outside exposed to all weathers (see "Breaking Seed Dormancy" page 233, for more information). Germination takes between 10 days and 6 months, but can be erratic, so be patient and do not discard the container. When the seeds have germinated, move young plants to a cold frame. Can flower in its first year.

OR

Sow in early autumn in prepared open ground. Mark the row or area clearly. Germination can take from 10 days to the following spring.

***Recipe for sowing, planting, and growing* Angelica archangelica**
Ingredients
5 seeds per cell or 8 seeds per pot
1 flat with cell packs
OR
1 x 4 in. pot
Standard loam-based seed mix mixed with coarse horticultural sand to a ratio of 1 part soil mix + 1 part sand
White plastic plant label
OR
Prepared site in the garden
2 white plastic plant labels

Method In early autumn, fill the cell packs or pot with soil mix, smooth over, tap down and water in well. Use fresh seed, sow 8 seeds per pot or 5 seeds per cell. Having sown the seeds in the pots, cover lightly with soild mix. Label with the plant name and date. Place the pot or cell pack outside, on a level surface, so that it is exposed to all weathers, including frosts. Do not worry if you live in a snowy area and the containers get immersed in snow, for melting snow will aid germination. If you live in an area that will not get a winter frost, it is a good idea to put the seed and a handful of damp sand in a plastic bag, which is marked clearly. Seal the bag. Place in the refrigerator for 3 weeks. Remove and sow as described above, then place outside.

Whichever method you use, germination is a bit erratic, taking anything from 10 days to 6 months. Do not give up and discard your container, the seeds might germinate next week. When the seeds have germinated, move young plants to a cold frame. Angelica can flower in its first year.

Alternatively, you can sow directly in a prepared site in the garden in early autumn. Space the seeds 12–18 in. apart, press gently into the soil and cover lightly, water in well. Mark the area clearly with 2 labels. Germination can take from 10 days, or not until the following spring.

Anthriscus *Apiaceae*

A genus of annual, biennial, and perennial herbs. Clusters of a small, white umbels of flowers in summer. Plant in a light soil with a degree of moisture retention and in semishade.

Anthriscus cerefolium (Chervil) all zones
Culinary herb
Medium seeds: 1,190 per ounce
Hardy annual (occasional biennial)
Height 1–2 ft.
Clusters of tiny white flowers
Flowers in summer
Light green aromatic, fern-like leaves
One of the traditional *fines herbes* in French cuisine

Sow fresh seeds in early spring in pots using standard soil-less seed mix, either peat or a peat substitute. Cover with perlite or vermiculite, place under protection at 60°F. Germination takes 5–10 days.
OR
Sow seeds in late spring in prepared open ground when the air temperature does not go below 45°F at night. Germination takes 2–3 weeks.

Apium *Apiaceae*

A genus of annual, biennial, or perennial herbs. They all have white flowers. Plant in semishade with a damp soil for best results.

Apium graveolens (Celery leaf)
Culinary herb
Small seeds: 3,200 per ounce
Hardy biennial grown as an annual
Height 1–3 ft.
Tiny, green-white flowers
Flowers in its second summer
Cut, midgreen leaves

Sow fresh seeds in early spring in pots using standard soil-less seed mix, either peat or a peat substitute. Cover with perlite or vermiculite, place under protection at 60°F. Germination takes 5–10 days.
OR
Sow seeds in late spring in prepared open ground, when the air temperature does not go below 45°F at night. Germination takes 2–3 weeks.

Arnica *Asteraceae*

A genus of perennial, rhizomatous herbs. They are all hardy and have yellow, daisy-like flowers throughout the summer. Plant in free-draining, humus-rich soil in a sunny situation. Both the species mentioned are good in rock gardens.

Arnica chamissonis (North American arnica) Zones 6–9
Medicinal herb
Small seeds: 5,040 per ounce
Hardy perennial
Height 20 in.
Clusters of yellow, daisy-like flowers

Apium graveolens

Flowers throughout the summer
Oval, light green, slightly hairy leaves

Arnica montana (Leopard's bane) Zones 6–9
Medicinal herb
Medium seeds: 2,520 per ounce
Hardy perennial
Height 12 in.
Large, single, scented yellow flowers
Flowers throughout the summer
Oval, light green, hairy leaves

Sow fresh seed in autumn in pots or cell packs, using standard loam-based mix. Cover lightly with soil mix, place in a cold frame. Germination takes 3–4 weeks. If no germination occurs during this time, place the container outside exposed to all weathers (see "Breaking Seed Dormancy", page 233, for more information). Germination can take a further 5–7 months on average, or even as long as 2 years. When germination has taken place, overwinter young plants in the cold frame. Can flower in its first year.

Artemisia *Asteraceae*

A very large genus of perennials, annuals, and shrubs, some of which are evergreen or semievergreen. Nearly all have aromatic foliage. Usually the flowers are small and insignificant; this is compensated by the foliage, which, in many cases, is a very attractive silver. Plant in a sunny situation in a well-drained soil.

Artemisia absinthium (Wormwood) Zones 4–9
Medicinal herb
Tiny seeds: 42,000 per ounce
Hardy perennial
Height 3 ft.
Tiny, insignificant, yellow flower heads borne in sprays
Flowers in summer
Abundant, divided, aromatic, gray green leaves

Artemisia annua (Sweet Annie)
Medicinal herb
Minute seeds: 84,000 per ounce
Hardy annual
Height 5 ft.
Tiny, yellow flowers in loose panicles
Flowers in summer
Highly aromatic, pinnate, divided, saw-toothed leaves

Artemisia dracunculoides (Russian tarragon) Zones 5–9
Culinary herb
Small seeds: 16,800 per ounce
Hardy perennial
Height 3 ft.
Tiny, insignificant, yellow flower heads borne in sprays
Flowers in summer
Aromatic, long, narrow, slightly coarse, green leaves

Artemisia vulgaris (Mugwort) Zones 4–9
Medicinal herb

Small seeds: 25,200 per ounce
Hardy perennial
Height 4 ft.
Panicles of insignificant, red brown flowers
Flowers in late summer
Pinnate, aromatic, dark green leaves with white undersides

Sow fresh seeds in spring in pots or cell packs using standard soil-less seed mix, either peat or a peat substitute. As these are very fine seeds, mix with the finest sand or talcum powder for an even sowing. Do not cover. Water from the bottom or with a fine spray. Place under protection at 68°F. Germination takes 10–20 days.

Borago *Boraginaceae*

A genus of annual and perennial herbs, native to the Mediterranean region. They all have attractive flowers. Plant in a sunny situation and a well-drained fertile soil. These plants can self-seed and can be invasive.

Borago officinalis (Borage)
Culinary and medicinal herb
Medium seeds: 168 per ounce
Hardy annual
Height 2 ft.
Loose racemes of blue-purple, star-shaped flowers
Flowers in summer
Oval to oblong, bristly, light-green leaves

Borago officinalis "Alba" (White borage)
Culinary herb
Medium seeds: 168 per ounce
Hardy annual
Height 20 in.
Loose racemes of white, star-shaped flowers
Flowers in summer
Oval to oblong, bristly, light gray green leaves

Sow seeds in early spring in pots using standard soil-less seed mix, either peat or a peat substitute. Cover with perlite or vermiculite, place under protection at 68°F. Germination takes 5–14 days.
OR
Sow seeds in late spring in prepared open ground, when the air temperature does not go below 45°F at night. Germination takes 2–3 weeks.

Calendula *Asteraceae*

A genus of shrubs and annuals. The shrubs require a minimum night temperature of 41°F. The annuals are hardy. Both should be planted in a sunny situation in a free-draining soil.

Calendula officinalis (Pot marigold)
Culinary and medicinal herb
Medium seeds: 420 per ounce
Hardy annual
Height 18 in.

Daisy-like single yellow or orange flowers
Flowers from spring to autumn
Light green, slightly aromatic lance-shaped leaves

Calendula officinalis 'Fiesta Gitana' (Marigold 'Fiesta Gitana')
Culinary and medicinal herb
Medium seeds: 140 per gram
Hardy annual
Height 20cm (8in)
Daisy-like double yellow or orange flowers
Flowers from spring to autumn
Light green, slightly aromatic, lance-shaped leaves

Sow seeds in early spring into pots using standard soil-less seed compost (substrate), either peat or a peat substitute. Cover with perlite or vermiculite, place under protection at 20°C (68°F). Germination takes 5–14 days.
OR
Sow seeds in late spring into prepared open ground, when the air temperature does not go below 5°C (41°F) at night. Germination takes 2–3 weeks.

Calomeria *Asteraceae*

There is only one species in the genus Calomeria which is native to Australia. Plant in well-drained soil in sun or partial shade. In cooler climates it makes a spectacular container plant.

Calomeria amaranthoides (*Humea elegans*) (Incense plant, Plume humea, Plume bush) DANGEROUS IRRITANT
Aromatic herb
Medium seeds: 450 per gram
Half-hardy annual in cool climates. (Biennial or short-lived perennial in warmer climates)
Height 3m (9ft)
Tiny delicate coral bracts with minute cream daisy-like flowers
Flowers from summer until early autumn
Highly aromatic, mid-green, large, oblong-shaped leaves

Sow seeds in autumn into pots or modules, using standard soil-less seed compost (substrate), either peat or a peat substitute, mixed with extra perlite or vermiculite for aeration. Mix to a ratio of 3 parts compost + 1 part perlite or vermiculite. Do not cover the seeds, place in a cold frame or unheated greenhouse. Germination takes 7–9 months, but can be erratic.

Carum *Apiaceae*

A genus of annual, biennial or perennial plants which have white or pink flowers displayed in compound umbels. Plant in a sunny situation and a well-drained, fertile soil.

Carum carvi (Caraway)
Culinary and medicinal herb
Medium seeds: 300 per gram
Hardy biennial

Height 20cm (8in) in the first year, 60cm (2ft) in the second year
Clusters of small umbels of pinkish-white flowers
Flowers in its second summer
Flowers followed by oblong aromatic seeds
Leaves feathery, light green, similar to carrot leaves

Sow fresh seeds in early spring into pots using standard soil-less seed compost (substrate), either peat or a peat substitute. Cover with perlite or vermiculite, place under protection at 15°C (60°F). Germination takes 5–10 days.
OR
Sow fresh seeds in late spring into prepared open ground, when the air temperature does not go below 7°C (45°F) at night. Germination takes 2–3 weeks.
    Sowing directly into open ground gives the best crop, because caraway hates being transplanted.

Cedronella *Lamiaceae*

This genus contains the single species *Cedronella canariensis* described below. Plant in a sunny position in a well-drained sandy loam.

Cedronella canariensis (Balm of Gilead)
Medicinal herb
Medium seed: 900 per gram
Half-hardy perennial, partial evergreen
Height 1m (3ft)
Aromatic pale mauve/pink 2-lipped flowers
Flowers throughout the summer
Strongly eucalyptus-scented mid-green trifoliate leaves

Sow seeds in spring into pots using standard soil-less seed compost (substrate), either peat or a peat substitute. Cover with perlite or vermiculite, place under protection at 20°C (68°F). Germination takes 14–20 days but can be spasmodic, so be patient.

Chamaemelum *Asteraceae*

A genus of annuals and evergreen perennials which have daisy-like flowers in summer. Plant in a sunny position in a fairly dry, light rich soil.

Chamaemelum nobile (Roman chamomile)
Medicinal herb
Small seeds: 6,500 per gram
Hardy perennial
Height 30cm (1 1/2ft)
White daisy-like flowers with yellow centres
Flowers all summer
Sweet-smelling finely divided mid-green foliage

Sow seeds in early spring into pots using standard soil-less seed compost (substrate), either peat or a peat substitute. Cover with perlite or vermiculite, place under protection at 18°C (65°F). Germination takes 14–20 days.

Chenopodium *Chenopodiaceae*

A genus of about 150 species of herbaceous subshrubs and annuals. The flowers are small and insignificant. The leaves and the seeds are both important as vegetable crops. Plant in sun or semishade in any alkaline soil. These plants are tolerant of salt.

Chenopodium ambrosioides (Epazote)
Culinary and medicinal herb
Small seeds: 18,900 per ounce
Half-hardy annual
Height 4 ft.
Tiny, greenish-brown flowers
Flowers in early summer
Arrow-shaped green leaves with a tinge of red

Chenopodium bonus-henricus (Good King Henry) Zones 5–9
Culinary herb
Medium seeds: 1,176 per ounce
Hardy perennial
Height 2 ft.
Tiny greenish-yellow flowers
Flowers in early summer
Green, arrow-shaped leaves

Sow seeds in early spring in pots using standard soil-less seed mix, either peat or a peat substitute. Cover with perlite or vermiculite, place under protection at 65°F. Germination takes 14–20 days.
OR
Sow seeds in late spring in prepared open ground, when the air temperature does not go below 41°F at night. Germination takes 2–3 weeks.

Coriandrum *Apiaceae*

A genus of annuals. Plant in a sunny position in a well-drained soil.

Coriandrum sativum (Coriander, Cilantro)
Culinary and medicinal herb
Medium seeds: 392 per ounce
Tender annual
Height 2 ft.
Flowers white with a hint of pink
Flowers in summer
Lobed and pinnate aromatic green leaves

Sow seeds in early spring in pots using standard soil-less seed mix, either peat or a peat substitute. Cover with perlite or vermiculite, place under protection at 65°F. Germination takes 5–10 days.
OR
Sow seeds in late spring in prepared open ground, when the air temperature does not go below 45°F at night. Germination takes 2–3 weeks.

***Recipe for sowing, planting, and growing* Coriandrum sativum**

Coriander is a very useful herb for the kitchen, combining well with many dishes from curries to salads. It is one plant that hates being transplanted and, when it is, it feels threatened and bolts before going to seed. The following methods should give a good crop of leaves before the plant runs to flower.

Ingredients
10–15 seeds
1 x 5 in. pot
Standard soil-less seed mix, either peat or a peat substitute
Fine-grade perlite (wetted) or vermiculite
White plastic label
OR
Prepared site in the garden
2 white plastic labels

Method Fill the pot with the soil-less mix, firm and water in well. Sow the seeds thinly on the top of the mix, press in gently with the palm of your hand, cover with perlite, label with plant name and date. Place the pot in a warm, light place 65°F, not full sun. Keep watering to a minimum until germination has taken place, which takes 5–10 days in late spring. Once the seedlings start germinating, make sure the container gets as much light as possible. If you live in a mild climate, where air temperature does not go below 45°F, place the container outside during the day, bringing it in at night. Continue until the third leaf starts to appear, then the container can be left out all night. If you are keeping the pot on the window sill, rotate daily so that the plant does not start growing towards the light. Start picking the leaves once they are large enough; this will encourage the new growth to develop. Start a second pot 4 weeks later to ensure a continuous crop.

Alternatively, you can sow directly in a prepared site in the garden when the night-time temperature does not fall below 50°F. Space the seeds 1 in. apart in a drill 1 in. deep. Cover lightly with soil and water in well. Label either end of the seed row. Germination takes 10–20 days. Start picking as soon as the leaves are large enough. Start a second row 4 weeks later to ensure a continuous crop.

Cuminum *Apiaceae*

A genus of half-hardy annuals. Plant in a sunny situation in a free-draining soil. The plant needs a minimum of 60°F to grow.

Cuminum cyminum (Cumin)
Culinary and medicinal herb
Medium seeds: 840 per ounce
Half-hardy annual
Height 6–12 in.
Umbels of white or pinkish flowers
Flowers in summer
Dark-green, finely divided leaves

Sow seeds in spring in pots using standard soil-less seed mix, either peat or a peat substitute. Cover with perlite or vermiculite, place under protection at 65°F. Germination takes 5–10 days.

Cymbopogon *Poaceae*

A genus of half-hardy perennial, aromatic grasses, mostly tropical. These plants can be grown in pots in a temperature no lower than 45°F. Keep watering to a minimum in winter.

Cymbopogon citratus
Culinary and medicinal herb
Small seeds: 16,600 per ounce
Half-hardy perennial
Height 3 ft.
Flowers rarely in cultivation and only in the tropics
Lemon-scented linear leaves

Sow seeds in early spring in pots using standard soil-less seed mix, either peat or a peat substitute. Cover with perlite or vermiculite, place under protection at 68°F. Germination takes 15–25 days.

Cynara *Asteraceae*

A genus of large, thistle-like perennials. Lovely, large, blue, violet, or white flower heads. Plant in a sunny situation in a well-drained fertile soil.

Cynara cardunculus (Cardoon) Zones 6–9
Culinary herb
Medium seeds: 67 per ounce
Hardy perennial
Height 3–8 ft.
Dramatic, multi-branched thistle heads and lovely purple flowers
Flowers in midsummer
Handsome, gray blue leaves

Sow seeds in early spring in pots using standard soil-less seed mix, either peat or a peat substitute. Cover with perlite or vermiculite, place under protection at 65°F. Germination takes 5–10 days.

Cynara cardunculus

OR
Sow seeds in late spring in prepared open ground, when the air temperature does not go below 45°F at night. Germination takes 2–3 weeks.

Diplotaxis *Brassicaceae*
A genus of annuals and perennials native to the Mediterranean area. For the best crop, sow directly into the garden, in a lightly shaded position in a rich moist soil.

Diplotaxis muralis (Rocket, Wild rocket) Zones 6–9
Culinary herb
Small seeds: 12,600 per ounce
Perennial
Height 12 in.
Yellow flowers
Flowers in summer
Green, serrated and toothed leaves, very strongly flavored

Sow seeds in early spring in pots using standard soil-less seed mix, either peat or a peat substitute. Cover with perlite or vermiculite, place under protection at 65°F. Germination takes 5–10 days.
OR
Sow seeds in late spring in prepared open ground, when the air temperature does not go below 45°F at night. Germination takes 2–3 weeks.

Dipsacus *Dipsacaceae*
A genus of hardy, erect, hairy or prickly biennial plants. Plant in sun or semishade in a damp soil (clay).

Dipsacus fullonum subsp. *fullonum* (Teasel) Zones 3–9
Medicinal herb
Medium seed: 980 per ounce
Hardy biennial
Height up to 6 ft.
Small mauve flowers, which appear in successive circles around the cone-shaped, spiny flower head
Flowers in summer
Bristly green leaves

Sow fresh seeds in autumn in pots or cell packs using standard soil-less seed mix, either peat or a peat substitute. Cover with soil-less mix, place in a cold frame. Germination takes 3 weeks to 4 months and can be erratic. Overwinter young plants in the cold frame. Flowers in second year.

Echinacea *Asteraceae*
A genus of perennial plants native to the United States. All have lovely, lightly scented, daisy-like flowers with conical centers. Plant in a sunny situation in a rich, free-draining soil.

Echinacea pallida (Pale Cone Flower) Zones 4–8
Medicinal herb
Medium seeds: 504 per ounce
Hardy perennial
Height 4 ft.
Purple, honey-scented, daisy-like flowers with conical orange brown centers
Flowers from summer to early autumn
Lance-shaped, midgreen leaves

Sow seeds in early spring in pots using standard soil-less seed mix, either peat or a peat substitute. Cover with perlite or vermiculite, place under protection at 65°F. Germination takes 15–24 days. If no germination has occurred after 28 days, it is worth putting the container outside for 21 days then bring back into the heat for a further 14 days. It can flower in the first year, but is more likely to flower in the second.

***Recipe for sowing, planting, and growing* Echinacea pallida**
Echinacea has become a very important medicinal herb. The part used in medicine is the root, so you will need a few plants if you are growing it for this purpose. However, the flower alone makes echinacea worth growing, it looks lovely in any bed.

Ingredients
5 seeds per cell or 9 seeds per pot
1 flat with cell packs
OR
1 x 4 in. pot
Standard soil-less seed mix, either peat or a peat substitute
Fine-grade perlite (wetted) or vermiculite
White plastic plant label

Method In early spring, fill the cell packs or pot with soil-less mix, smooth over, tap down and water in well. Sow the seeds thinly on the surface of the soil-less mix. Cover with perlite or vermiculite and label with the plant name and date. Place the flat or pot in a warm, light place, out of direct sunlight, at an optimum temperature of 65°F. Keep watering to a minimum until germination has taken place, which takes 15–24 days. If no germination has occurred after 28 days, it is worth putting the container outside for 21 days, then bring back into its warm, light place until germination starts. As soon as you see the seedlings emerge, remove the container from the heat, place in a cooler, warm light place and grow on until the seedlings are large enough to handle, approximately 3–4 weeks after germination. Either prick out and pot up, or. if you are using cell packs. you can plant out in the garden after a period of hardening off and when there is no threat of frost.

Echium *Boraginaceae*
A genus of annuals, biennials, perennials, and evergreen shrubs, some of which are tender. All grown for their lovely flowers. Plant in a sunny site in a well-drained fertile soil.

Echium vulgare (Viper's bugloss)
Medicinal herb
Medium seeds: 644 per ounce
Hardy biennial often grown as an annual
Height 20 in.
Lovely flowers, purple in bud, then blue and pink when in flower
Flowers in early summer
Bristly oval leaves

Sow fresh seeds in autumn in pots or cell packs using a standard loam-based seed mix mixed with coarse horticultural sand. Mix to a ratio of 1 part soil mix + 1 part sand. Cover with coarse horticultural sand, then place outside exposed to all weathers. (see "Breaking Seed Dormancy", page 233, for more information). Germination takes 3 weeks to 4 months, but can be erratic, so be patient, and do not discard the container. Flowers in its second year.

Eruca *Brassicaceae*

A genus of annuals and perennials native to the Mediterranean area. For the best crop, sow directly in a lightly shaded position in a rich, moist soil.

Eruca vesicaria subsp. *sativa* (Rocket, Salad rocket, Arugola, Rucola)
Culinary Herb
Medium seeds: 1,960 per ounce
Annual
Height 2–3 ft.
Cream flowers with purple veins
Flowers in summer
Green, toothed leaves, tasting of mustard, pepper, and beef

Sow seeds in early spring in pots using standard soil-less seed mix, either peat or a peat substitute. Cover with perlite or vermiculite, place under protection at 65°F. Germination takes 5–10 days.
OR
Sow seeds in late spring in prepared open ground when the air temperature does not go below 45°F at night. Germination takes 2–3 weeks.

Eupatorium *Asteraceae*

A genus of hardy and half-hardy shrubs and perennials, grown mainly for their flowers. Plant in moist, free-draining soil in sun or partial shade.

Eupatorium cannabinum (Hemp Agrimony, Thoroughwort)
Zones 4–9
Medicinal herb
Small seeds: 7,840 per ounce
Perennial
Height 1–4 ft.
Clusters of mauve-pink flowers
Flowers from summer to early autumn
Green leaves divided into 3 or 5 segments

Eupatorium purpureum (Joe Pye weed) Zones 3–9
Medicinal herb
Small seeds: 10.640 per ounce
Perennial
Height 6–10 ft.
Clusters of pink flowers
Flowers in early autumn
Finely toothed, oval leaves with a faint vanilla scent when crushed

Sow fresh seeds in autumn in pots or cell packs, using standard loam-based seed mix. Cover with fine sand and place in a cold frame. Germination takes 6–10 weeks. If no germination has occurred after 28 days it is worth putting the container outside for 21 days, then back in the cold frame (see "Breaking Seed Dormancy", page 233, for more details). Overwinter young plants in a cold frame.

Filipendula *Rosaceae*

A genus of hardy perennials, most of which are good marginal water plants, flowering from midspring to midsummer. Plant in moist or boggy soil in sun or partial shade.

Filipendula ulmaria (Meadowsweet) Zones 3–9
Medicinal herb
Small seeds: 3,360 per ounce
Hardy perennial
Height 3 ft.
Frothy, creamy-white scented flowers
Flowers in midsummer
Darkish green, pinnate, serrated leaves, aromatic when crushed

Sow seeds in early spring in pots or cell packs using standard soil-less seed mix, either peat or a peat substitute. Cover with perlite or vermiculite and then place in a cold frame. Germination takes 1–3 months.

Foeniculum *Apiaceae*

This genus contains the single species *Foeniculum vulgare* described below. Plant in a sunny position in a well-drained, sandy loam, dislikes damp cold winters.

Foeniculum vulgare (Fennel) Zones 5–9
Culinary and medicinal herb
Medium seeds: 644 per ounce
Perennial sometimes biennial
Height 4¹/₂ ft.
Umbels of small yellow flowers, which are followed by gray brown aromatic seeds
Flowers in summer
Soft, green, feathery, aromatic foliage

Sow in spring in pots or cell packs, using standard soil-less seed mix, either peat or a peat substitute. Place under protection at 68°F. Germination takes 7–10 days.
OR
Sow in late spring in prepared open ground, when air temperature does not go below 41°F at night. Germination takes 14–20 days.

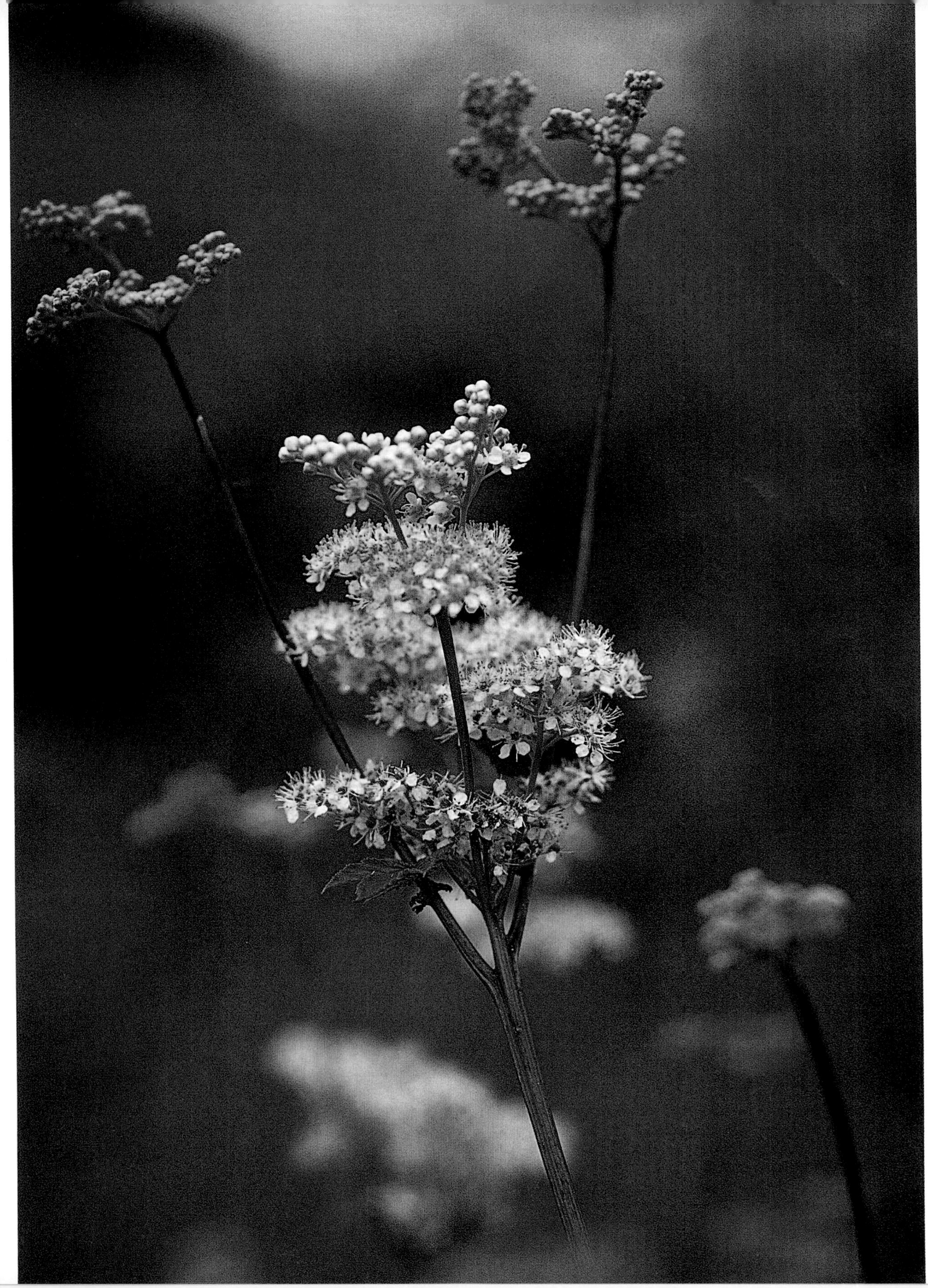

Filipendula ulmaria

Marrubium *Lamiaceae*

A genus of annuals and perennials. Plant in a well-drained soil in a sunny position.

Marrubium vulgare (Horehound) Zones 4–8
Medicinal herb
Small seeds: 3,080 per ounce
Hardy perennial
Height 18 in.
Small creamy flowers
Flowers in summer
Aromatic, downy, oval, gray-green leaves

Sow seeds in early spring in pots or cell packs using standard soil-less seed mix, either peat or a peat substitute. Cover with perlite or vermiculite, place under protection at 65°F. Germination takes 1–2 weeks; however, some years it can take 3–4 weeks and be a bit spasmodic.

Melissa *Lamiaceae*

A genus of half-hardy and hardy perennials. Plant in a well-drained soil in a sunny situation.

Melissa officinalis (Lemon balm) Zones 4–9
Culinary and medicinal herb
Small seeds: 4,480 per ounce
Hardy perennial
Height 2¹/₂ ft.
Clusters of small pale yellow white flowers
Flowers in summer
Oval, toothed, slightly wrinkled green leaves, which smell of lemon when crushed

Sow seeds in early spring in pots or cell packs, using standard soil-less seed mix, either peat or a peat substitute. Cover with perlite or vermiculite, place under protection at 68°F. Germination takes 1–2 weeks; however, some years it can take 3–4 weeks and can be spasmodic. Lemon balm can be invasive. It will adapt to all soils, with the exception of waterlogged sites.

Mentha *Lamiaceae*

A genus of hardy perennials, some of which are semievergreen. Plant in a well-drained soil, in sun or partial shade. Mentha is renowned for being invasive, so plant with care.

Mentha pulegium "Upright" (Pennyroyal, Pennyroyal Upright) Zones 6–9
Culinary and medicinal herb
Tiny seeds: 35,000 per ounce
Hardy perennial, semievergreen
Height 12 in.
Small mauve flowers
Flowers in late spring
Small, bright green leaves, which are very strongly peppermint scented

Sow seeds in spring in pots or cell packs using standard soil-less seed mix, either peat or a peat substitute. As these are very fine seeds, mix with the finest sand or talcum powder for an even sowing. Do not cover. Water from the bottom or with a fine spray. Place under protection at 68°F. Germination takes 10–20 days.

Micromeria *Lamiaceae*

A genus of shrubs and perennials, some of which are evergreen. Plant in a well-drained soil in a sunny situation.

Micromeria. (Emperor's mint) Zones 7–9
Culinary and medicinal herb
Small seeds: 8,400 per ounce
Hardy perennial
Height 12 in.
Small, pale pink gray flowers
Flowers in summer
Small, oval, pointed gray leaves, which have a strong mint aroma and flavor

Sow seeds in early spring in pots or cell packs using standard soil-less seed mix, either peat or a peat substitute. Cover with perlite or vermiculite, place under protection at 65°F. Germination takes 1–2 weeks.

Monarda *Lamiaceae*

A genus of annuals and perennials, all of which have aromatic foliage and attractive flowers. Plant in a light well-drained soil in a sunny or partially shaded position.

Monarda citriodora (Lemon bergamot)
Culinary and medicinal herb
Small seeds: 5,180 per ounce
Hardy annual
Height 2–2¹/₂ ft.
Stunning lavender flowers that come in 2 tiers
Flowers all summer
Oval, pointed, bright green leaves that have a minty lemon scent when crushed

Monarda fistulosa (Wild bergamot) Zones 3–9
Culinary and medicinal herb
Small seeds: 8,960 per ounce
Hardy perennial
Height 3–4 ft.
Attractive mauve flowers
Flowers in summer
Oval, pointed, aromatic, midgreen leaves

Sow seeds in early spring in pots or cell packs using standard soil-less seed mix, either peat or a peat substitute. Cover with perlite or vermiculite, place under protection at 65°F. Germination takes 1–2 weeks.

Melissa officinalis

Myrrhis *Apiaceae*

This genus contains the single species *Myrrhis odorata* described below. Plant in any soil, other than waterlogged, in sun or partial shade. In light soils this plant can be invasive.

Myrrhis odorata (**Sweet cicely**)
Culinary herb
Medium seeds: 22 per gram
Hardy perennial
Height 90cm (36in)
Large umbels of small white sweetly-scented flowers
Flowers in spring
Fern-like, very divided, bright green leaves which smell of aniseed when crushed

Sow seeds in autumn into pots or modules using a standard loam-based seed compost (substrate). Cover with coarse horticultural sand and place outside exposed to all the weathers (see 'Breaking Seed Dormancy', page 233, for more information). It is most important that you use a loam-based compost, this seed does not like peat. Germination takes 4–6 months, can be erratic, so be patient, and do not discard container.

Nepeta *Lamiaceae*

A genus of hardy perennials. Plant in a well-drained soil in a sunny position.

Nepeta cataria (**Catnip, Dog Mint, Nep-in-a-hedge**)
Medicinal herb
Small seeds: 1,700 per gram
Hardy perennial
Height 1m (3ft)
White to pale pink small flowers
Flowers from early summer to early autumn
Pungent, aromatic, oval, toothed leaves which the cats adore

Nepeta racemosa (Catmint) Zones 3–8
Medicinal herb
Small seeds: 3,640 per ounce
Hardy perennial
Height 20 in.
Spikes of lavender-blue purple small flowers
Flowers from late spring to autumn
Small, mildly fragrant, grayish-green leaves

Sow seeds in early spring in pots or cell packs using standard soil-less seed mix, either peat or a peat substitute. Cover with perlite or vermiculite, place under protection at 65°F. Germination takes 1–2 weeks.

Nigella *Ranunculaceae*

A genus of annuals with attractive flowers and seed heads. Plant in a free-draining soil in a sunny situation.

Nigella sativa (Black cumin, Nutmeg flower, Fennel flower)
Culinary herb
Medium seeds: 1,204 per ounce
Hardy annual
Height 12 in.
Small, very pale blue flowers
Flowers in summer
Finely divided leaves

Sow seeds in early spring in pots or cell packs using standard soil-less seed mix, either peat or a peat substitute. Cover with perlite or vermiculite, place under protection at 65°F. Germination usually takes 1–2 weeks, but can take 3–4 weeks and be a bit spasmodic.

Ocimum *Lamiaceae*

This is a genus of annuals, perennials, and shrubs, native of the tropics and especially Africa. Plant in a sunny, well-sheltered position in fertile, free-draining soil. Protect plants where nighttime temperatures drop below 48°F.

Ocimum basilicum (Sweet basil, Genovese basil)
Culinary herb
Medium seeds: 1,960 per ounce
Half-hardy annual
Height 18 in.
Whorls of small, white tubular flowers
Flowers in summer
Oval, green, sweetly scented and strongly flavored leaves

Ocimum tenuiflorum (Holy basil, Sacred basil, Tulsi)
Culinary and medicinal herb
Small seeds: 4,200 per ounce
Half-hardy annual
Height up to 12 in.
Small, mauve, violet, white flowers in long slender racemes
Flowers in summer
Slightly hairy, green oval leaves with magenta edges, which are spicy and pungent

Sow seeds in early spring in pots or cell packs, using standard soil-less seed mix, either peat or a peat substitute. Cover with perlite or vermiculite, place under protection at 68°F. Germination takes 1–2 weeks.

> ### Recipe for sowing, planting, and growing Ocimum
>
> Basil is the Rolls Royce of the kitchen. There are now many varieties available, all of them with unique flavors. It is a warm-climate plant and growing basil in cool climates can be problematic. The important thing is not to be in too much of a hurry and start the seeds off too soon; they hate fluctuating temperatures and cold nights. So, wait until spring has truly sprung.
>
> #### Ingredients
> 6–8 seeds per cell or 15 seeds per pot
> 1 flat with cell packs
> **OR**
> 1 x 4 in. pot
> Standard soil-less seed mix, either peat or a peat substitute
> Fine-grade perlite (wetted) or vermiculite
> White plastic plant label
>
> **Method** In early spring fill the cell packs or pot with soil-less mix, smooth over, tap down and water in well, sow the seeds thinly on the surface of the mix. Cover with perlite or vermiculite and label with the plant name and date. Place the flat or pot in warmth and light, out of direct sunlight, at an optimum temperature of 68°F. Keep watering, but do not overwater, until germination has taken place after 7–14 days. Once the seedlings start germinating, water only in the morning, before midday. Basil seedlings hate to be wet at night. It is also important to remove them from the extra heat as soon as you see the seedlings emerge. Place in a cooler, but warm, light place and grow on until the seedlings are large enough to handle, approximately 2–3 weeks after germination. Once the seedlings are large enough to handle, or the cell pack is rooted, remove from the flat and pot into a 4 in. pot. Water in, label, and allow to grow on, in a warm, well-lit environment until established. If you have started the seeds in a 4 in. pot, remove the seedlings from the pot, and split them gently, replanting 2 seedlings per pot, using a soil-less mix. Grow on for a further few weeks until the roots are emerging from the bottom of the pot. Plant out in the garden in a very sheltered position when all threat of frost has passed, or pot up in an attractive pot and keep near the kitchen. One word of warning: slugs love basil.

Oenothera *Onagraceae*

A genus of annuals, biennials, and perennials. Plant in a well-drained, preferably sandy soil, in a sunny position.

Oenothera biennis (Evening primrose) Zones 4–9
Medicinal herb
Small seeds: 7,000 per ounce
Hardy biennial
Height up to 3–5 ft.
Lovely, trumpet-shaped, night-scented, yellow flowers that are followed by oval, downy pods containing loads of small seeds
Flowers in summer
Oblong green leaves
This species can self-seed freely

Rosmarinus officinalis (Rosemary) Zones 7–9
Culinary and medicinal herb
Medium seeds: 2,100 per ounce
Evergreen hardy perennial
Height 3¹/₂ ft.
Pale blue flowers
Flowers in early spring until summer
Needle-shaped, dark green, highly aromatic leaves

Sow seeds in early spring in pots or cell packs using standard soil-less seed mix, either peat or a peat substitute. Cover with perlite or vermiculite, place under protection at 70°F. Germination takes 1–2 weeks. It is critical not to overwater after germination, as the seedlings are prone to damping off.

Ruta *Rutaceae*
A genus of hardy evergreen subshrubs. Plant in a well-drained soil in a sunny situation.

Ruta graveolens (Rue) Zones 4–9
DANGEROUS IRRITANT
Medicinal herb
Medium seeds: 1,680 per ounce
Evergreen hardy perennial
Height 2 ft.
Yellow waxy flowers with 4 or 5 petals
Flowers in summer
Small, rounded lobed leaves of a greeny-blue color

Sow seeds in early spring in pots or cell packs, using standard soil-less seed mix, either peat or a peat substitute. Cover with perlite or vermiculite, place under protection at 65°F Germination takes 1–2 weeks.

Salvia *Lamiaceae*
A genus of annuals, biennials, perennials, and evergreen shrubs and subshrubs. Plant in a fertile, well-drained soil in a sunny position.

Salvia officinalis (Sage) Zones 5–9
Culinary and medicinal herb
Medium seeds:392 per ounce
Evergreen hardy perennial
Height 2 ft.
Mauve/blue 2-lipped flowers
Flowers in summer
Textured, oval aromatic green leaves

Salvia sclarea (Clary sage, Muscatel sage) Zones 5–9
Culinary and medicinal herb
Medium seeds: 700 per ounce
Hardy biennial
Height up to 3 ft.
Small, pale violet flowers with bracts of blue, mauve, and white
Flowers in the second summer
Large, soft-green, slightly wrinkled leaves

Salvia viridis (Painted sage)
Culinary and medicinal herb
Medium seeds: 980 per ounce
Hardy annual
Height 18 in.
The most striking part of this flower is the colorful bracts, which are in shades of pink, purple, or white
Flowers in summer
Oval, midgreen leaves

Sow seeds in early spring in pots or cell packs, using standard soil-less seed mix, either peat or a peat substitute. Cover with perlite or vermiculite, place under protection at 65°F. Germination takes 1–2 weeks.

Sanguisorba *Rosaceae*
A genus of hardy perennials. Plant in a moist (not waterlogged) soil in a sunny or partially shaded position.

Sanguisorba minor (Salad burnet) Zones 5–9
Culinary herb
Medium seeds: 280 per ounce
Hardy perennial
Height 12 in.
Small spikes of dark crimson flowers
Flowers in early summer
Soft midgreen leaves, divided into oval leaflets with toothed edges

Sow fresh seeds in autumn in pots or cell packs, using standard soil-less seed mix, either peat or a peat substitute. Cover with perlite or vermiculite, place in a cold frame. Germination takes 2–3 weeks. Overwinter young plants in a cold frame.
OR
Sow seeds in early spring in pots or cell packs, using standard soil-less seed mix, either peat or a peat substitute. Cover with perlite or vermiculite, place under protection at 65°F. Germination takes 2–4 weeks.

Saponaria *Caryophyllaceae*
A genus of annuals and perennials. Plant in a well-drained soil in a sunny position.

Saponaria officinalis (Bouncing Bet) Zones 3–9
Medicinal herb
Medium seeds: 1,428 per ounce
Hardy perennial
Height up to 3 ft.
Compact clusters of small, pretty pink or white flowers
Flowers in late summer
Smooth oval midgreen leaves
This plant can be invasive

Sow fresh seed in autumn in pots or cell packs, using standard loam-based seed mix. Cover lightly with soil mix, place in a cold frame. Germination takes 3–4 weeks. If no germination occurs during this time, place the container outside exposed to all weathers (see

"Breaking Seed Dormancy", page 233, for more information). Germination can take a further 5–7 months on average, or even as long as 2 years. When germination has taken place, overwinter young plants in the cold frame.

***Recipe for sowing, planting, and growing* Saponaria officinalis**
This plant is a natural form of soap. The herb is also used medicinally and looks most attractive growing in the garden.

Ingredients
10 seeds per cell or 18 seeds per pot
1 flat with cell packs
OR
1 x 4 in. pot
Standard loam-based seed mix, mixed with coarse horticultural sand to a ratio of 2 parts soil mix + 1 part sand
White plastic plant label

Method In autumn, fill the cell packs or pot with standard, loam-based seed mix mixed with coarse horticultural sand. Smooth over, tap down and water in well. Sow the seeds thinly on the surface of the soil mix. Press gently into the mix with the palm of the hand. Cover the seed with coarse horticultural sand. Label with the plant name and date. Place the flat or pot in a cold frame.

Germination takes 3–4 weeks. If no germination has occurred after 28 days it is worth putting the container outside exposed to all weathers, including frosts. Do not worry if you live in a snowy area and the containers get immersed in snow, melting snow will aid germination (see "Breaking Seed Dormancy", page 233, for more information). Germination can take a further 5–7 months on average, or even as long as 2 years. Don't give up. When germination has taken place, overwinter young plants in the cold frame.

After germination, prick out the seedlings when they are large enough to handle. If you are using cell packs, you can plant directly in the garden as soon as the soil is warm enough to dig over before planting out.

Satureja *Lamiaceae*
A genus of annuals and semievergreen perennials. Plant in a well-drained soil in a sunny position.

Satureja hortensis (Summer savory)
Culinary herb
Small seeds:4,200 per ounce
Hardy annual
Height 12 in.
Small, white/mauve flowers
Flowers in summer
Highly aromatic, narrow, oblong midgreen leaves

Sow seeds in early spring in pots or cell packs using standard soilless seed mix, either peat or a peat substitute. Cover with perlite or vermiculite, place under protection at 68°F. Germination takes 1–2 weeks

Scutellaria *Lamiaceae*
A genus of hardy and half-hardy rhizomatous perennials. Plant in a well-drained soil in a sunny position.

Scutellaria galericulata (Skullcap) Zones 6–9
Medicinal herb
Small seeds: 4,200 per ounce
Hardy perennial
Height up to 20 in.
Small purple/blue flowers
Flowers in summer
Lance-shaped green leaves

Sow fresh seeds in autumn in pots or cell packs, using standard soilless seed mix, either peat or a peat substitute. Cover with perlite or vermiculite, place in a cold frame. Germination takes 3–4 weeks. If no germination occurs during this time, place the container outside exposed to all weathers (see "Breaking Seed Dormancy", page 233, for more information). Germination can take a further 5–7 months. Overwinter young plants in a cold frame. Can flower in its first season.

Solidago *Asteraceae*
A genus of hardy perennials. Can be grown in most soils in sun or partial shade.

Solidago virgaurea (Goldenrod) Zones 5–9
Medicinal herb
Small seeds: 3,360 per ounce
Hardy perennial
Height 18 in.
Small yellow flowers
Flowers in late summer
Small, lance-shaped, finely toothed midgreen leaves

Sow seeds in autumn in pots or cell packs using a standard loam-based seed mix, mixed with coarse horticultural sand. Mix to a ratio of 1 part soil mix + 1 part sand. Cover with coarse horticultural sand. Then place outside exposed to all the weathers (see "Breaking Seed Dormancy", page 233, for more information). Germination takes 4–6 months.

Tanacetum *Asteraceae*
A genus of hardy perennials, some of which are evergreen. Plant in a well-drained soil in a sunny position.

Tanacetum cinerariifolium (Pyrethrum) Zone 6–10
Medicinal herb
Medium seeds: 1,820 per ounce
Hardy perennial
Height 2¹/₂ ft.
White daisy-like flowers with yellow centers
Flowers in summer
Finely divided gray green leaves with white down on the underside

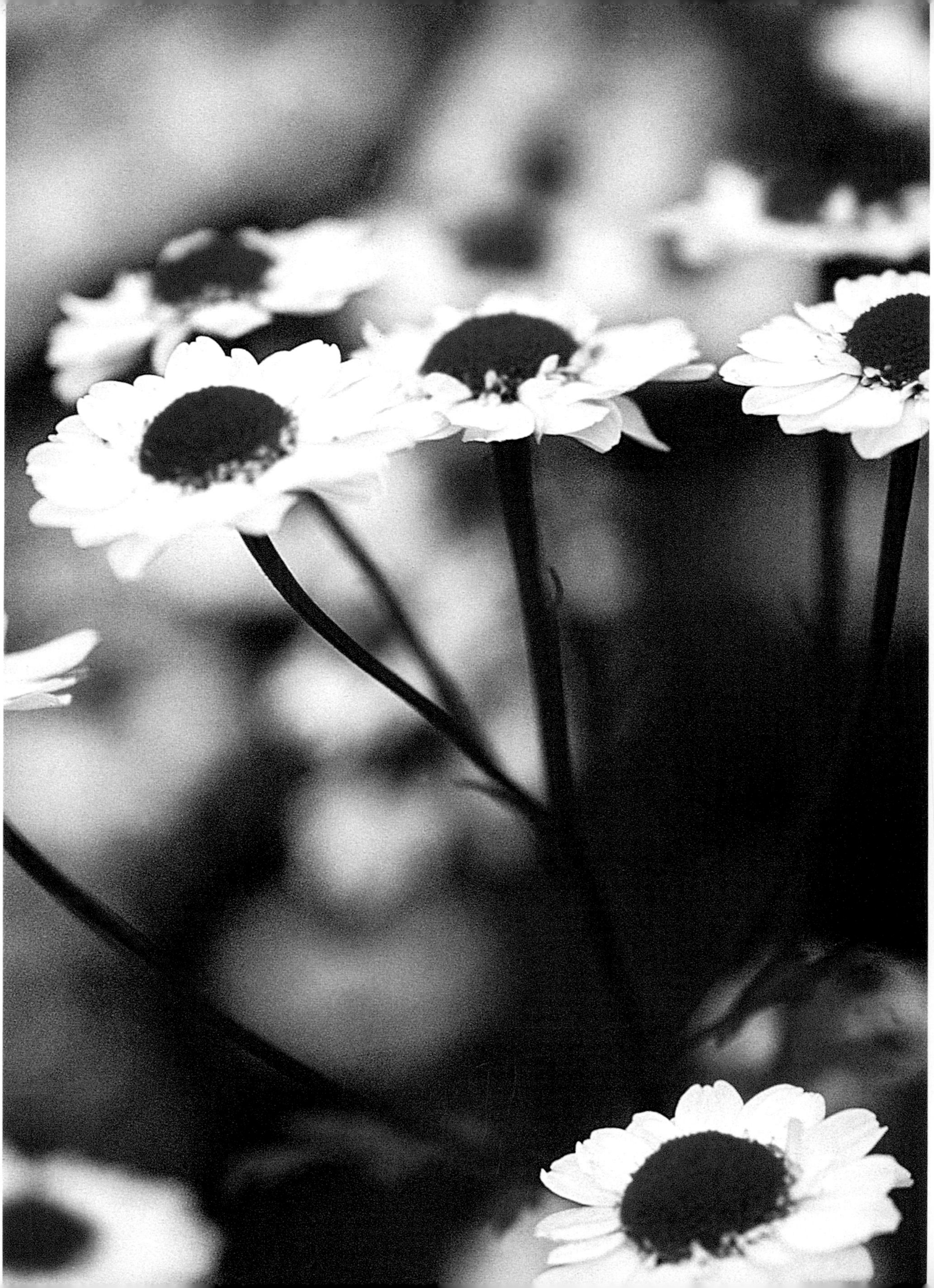

Tanacetum parthenium

Tanacetum parthenium (Feverfew) Zones 5–9
Culinary and medicinal herb
Small seeds: 25,200 per ounce
Hardy perennial
Height up to 18 in.
Small white daisy-like flowers with yellow centers
Typical chrysanthemum-shaped leaves, midgreen and divided
This plant self-seeds easily, so can be invasive

Tanacetum vulgare (Tansy) Zones 3–9
Culinary and medicinal herb
Small seeds: 23,800 per ounce
Hardy perennial
Height 3 ft.
Yellow button flowers in late summer
Deeply indented, toothed midgreen aromatic leaves
The roots of this plant can be invasive

Sow seeds in spring in pots or cell packs, using standard soil-less seed mix, either peat or a peat substitute. The small seeds should be mixed with the finest sand or talcum powder for an even sowing. Do not cover. Water from the bottom or with a fine spray, place in a cold frame. The mediumsized seeds should be sown on the surface of the soil-less mix, left uncovered and placed in a cold frame. Germination takes 2–3 weeks.

Teucrium *Lamiaceae*

A genus of perennials, evergreens, shrubs, and subshrubs. Plant in a well-drained soil in a sunny position.

Teucrium scorodonia (Wood sage) Zones 6–9
Medicinal herb
Small seeds: 4,200 per ounce
Hardy perennial
Height up to 12 in.
Pale greenish-white flowers
Flowers in summer
Soft green, heart-shaped leaves with a mild smell of garlic when crushed

Sow seeds in autumn in pots or cell packs, using a standard loam-based seed mix, mixed with coarse horticultural sand. Mix to a ratio of 1 part soil mix + 1 part sand. Cover with coarse horticultural sand. Then place outside exposed to all weathers (see "Breaking Seed Dormancy", page 233, for more information). Germination takes 4–6 months.

Thymus *Lamiaceae*

A genus of evergreen perennials. Plant in a well-drained soil in a sunny position.

Thymus serpyllum (Mother of thyme) Zones 4–9
Culinary and medicinal herb
Small seeds: 19,600 per ounce
Evergreen hardy perennial
Creeping habit
Small mauve/purple flowers
Flowers in summer
Small, dark green, oval aromatic leaves

Thymus vulgaris (Common thyme) Zone 5–9
Culinary and medicinal herb
Small seeds: 19,600 per ounce
Evergreen hardy perennial
Height 12 in.
Small mauve flowers
Flowers in summer
Small, oval, green, highly aromatic leaves

Sow seeds in spring in pots or cell packs, using standard soil-less seed mix, either peat or a peat substitute. As these are very fine seeds, mix with the finest sand or talcum powder for an even sowing. Do not cover. Water from the bottom or with a fine spray. Place under protection at 68°F. Germination takes 5–10 days.

Valeriana *Valerianaceae*

A genus of hardy perennials. Plant in any soil, including damp, in a sunny situation.

Valeriana officinalis (Valerian) Zones 5–9
Medicinal herb
Small seeds: 3,920 per ounce
Hardy perennial
Height 4 ft.
Pale pink/white flowers
Flowers in summer
Mid-green, deeply toothed leaves

Sow seeds in spring in pots or cell packs, using standard soil-less seed mix, either peat or a peat substitute. Cover with perlite or vermiculite and place in a cold frame. Germination takes 3–4 weeks.

Verbena *Verbenaceae*

A genus of hardy and half-hardy biennials and perennials. Plant in a well-drained soil in a sunny situation.

Verbena officinalis (Vervain) Zones 4–9
Medicinal herb
Small seeds: 3,920 per ounce
Hardy perennial
Height 2½ ft.
Very small, pale lilac flowers
Flowers in summer
Midgreen hairy leaves, which are often deeply divided into lobes with curved teeth

Sow seeds in autumn in pots or cell packs, using a standard loam-based seed mix, mixed with coarse horticultural sand to a ratio of 1 part soil mix + 1 part sand. Cover with coarse horticultural sand. Then place outside exposed to all weathers (see "Breaking Seed Dormancy", page 233, for more information). Germination takes 4–6 months, but can be erratic, so be patient and do not discard container.

Palms are lovely evergreen, tropical, temperate trees, which look graceful growing outside in warm and tropical climates. In cool climates, they make a very good architectural plant for a conservatory or hothouse.

Cycads have become very popular plants, they are among the most primitive living seed plants, often called 'living fossils', having changed very little in the last 200 million years. A 220-year-old specimen of *Encephalartos*, a relative of *Cycas revoluta*, is on display at the Royal Botanic Garden, Kew, England; the restoration of the famous Palm House meant it had to be transplanted temporarily to a holding area for more than a year; the move was successful and is an example of the durability of these ancient 'living fossils'. Cycads in cool climates make wonderful house plants as they are very tolerant, needing only a few hours of good light per day.

palms & cycads

Borassus *Palmaceae*

A genus of palms with one species described below. Plant in a sandy loam in a sunny position. In cool climates, grow as a large container plant.

Borassus flabellifer (Palmyra palm) Zone 10
Very large seed
Tropical evergreen tree
Height 80 ft.
Pale yellow flowers in winter, followed by large fruits each holding three seeds in spring
The leaves are nearly circular and pleated like a fan with about 40 ribs radiating from a common center

As this is a very large fruit that produces a long tap root, sow direct in a deep container in spring, using a standard soil-less seed mix, either coir or peat, mixed with 1/16–1/8 in. fine grit. Mix to a ratio of 1 part soil-less mix + 1 part fine grit. Half bury the seed and place under protection at 77°F. Germination takes 2–4 months.

Grow on as a container plant for 3 years, before planting out in warm climates. Alternatively, in cool climates grow on as a container plant, using a standard soil-less seed mix, either peat or peat substitute, mixed with extra fine potting bark for extra aeration. Mix to a ratio of 3 parts soil-less mix + 1 fine bark.

Caryota *Palmaceae*

This genus of palms consists of tropical evergreen trees. Plant in a rich, moist, well-drained soil in sun or partial shade. In cool climates grow as a large container plant.

Caryota mitis (Fish-Tail Palm) Zone 10
Large seeds
Tropical evergreen tree
Height up to 25 ft.
Clusters of cream flowers, followed by round, red to black fruit
Flowers in winter
The slender stem is topped with several bipinnate leaves that can reach 9 in. in length. The light-green leaflets are shaped like a fish's lower fin, hence its common name

In spring remove the flesh carefully from the seed (see warning below), then soak the seed for 24 hours and sow immediately in pots using standard soil-less seed mix, either coir or peat, mixed with 1/16–1/8 in. fine grit. Mix to a ratio of 3 parts soil-less mix + 1 part fine grit. Cover with fine grit, then place under protection at 68°F. Germination takes 1–3 months

Grow on as a container plant for 3 years, before planting out in warm climates. Alternatively, in cool climates, grow on as a container plant using a standard soil-less seed mix, either peat or peat substitute, mixed with extrafine potting bark for extra aeration. Mix to a ratio of 3 parts soil-less mix + 1 fine bark.

WARNING—Avoid contact with the red fruit produced by this palm. It contains oxalic acid, which is toxic when ingested, and contact with skin could result in severe chemical burns.

Cocus *Palmaceae*

A single genus of tropical evergreen palm tree, as described below. Plant in a sandy loam in full sun. In cool climates, grow as a container plant.

Cocus nucifera (Coconut) Zone 10
Very large seed
Tropical evergreen tree
Height up to 100 ft.
Loose clusters of cream flowers
Flowers in winter
Large, palm-shaped leaves with long, pinnate leaflets

As this is a very large fruit ,which produces a long tap root, in spring or summer soak the seed in water for 2 days, remove all the fiber (coir) from around the nut, scarify (see page 233) then sow direct in a deep container. Use standard soil-less seed mix, either coir or peat, mixed with 1/16–1/8 in. fine grit to a ratio of 1 part soil-less mix + 1 part fine grit. Place the seed on its side and half bury the seed. Then place under protection at 80°F. Germination takes 2–4 months

Grow on as a container plant for 2 years, before planting out in warm climates. Alternatively, in cool climates, grow on as a container plant using a standard soil-less seed mix, either peat or peat substitute, mixed with extra fine potting bark for extra aeration. Mix to a ratio of 3 parts soil-less mix + 1 part fine bark.

Cycas *Cycadaceae*

A genus of cycads that consists of tropical and subtropical evergreen shrubs and perennials. Plant in a well-drained soil, which is rich in humus, in a sunny position. In cool climates, this plant makes a very good houseplant or conservatory plant.

Cycas circinnalis (Fern Palm, Sago Palm) Zone 10
Large seeds
Tropical evergreen shrub
Height up to 12 ft.
Female cone-like structure, which bears nut-like seeds. A mature male and female cycad are needed to produce viable seeds.
A rugged trunk, topped with whorled, feathery leaves

In spring, soak the seed for 48 hours, discarding any floating seeds, then make a small nick in the end of the seed, and sow immediately in deep pots using standard soil-less seed mix, either coir or peat, mixed with coarse grit. Mix to a ratio of 3 parts soil-less mix + 1 part coarse grit. Position the seed sideways, with only the top edge exposed, then place under protection at 54°F in shade. Germination takes 3–9 months.

Grow on as a container plant for 3 years, before planting out in warm climates. Alternatively, in cool climates, grow on as a container plant using a standard soil-less seed mix, either peat or peat substitute, mixed with coarse grit for extra aeration. Mix to a ratio of 3 parts soil-less mix + 1 coarse grit

Caryota mitis

Recipe for sowing, planting, and growing Cycas circinnalis
Ingredients
1 seed per pot
Small bowl
Hot, not boiling, water
1 sharp knife
1 x 4 in. pot
Standard soil-less seed mix, either coir or peat, mixed with coarse grit
  to a ratio of 3 parts soil-les mix + 1 part grit
Waterproof pen
White plastic plant label
OR
1 plastic bag
Coir or sphagnum peat moss
Somewhere constantly warm

Method The seed develops over the summer and is ripe by late winter or early spring. Check to see if the seed is viable by shaking the seed: if you hear a rattle, discard the seed. With the remaining seed, remove the skin and drop seed into a bowl of water; if any seed floats, it is not viable. Before sowing in spring, soak seed in water for several days, then remove the skin and make a small nick, not too deep, with a sharp knife in the end of the seed and sow immediately. If you are using purchased, rather than fresh, seeds, soak the seeds in warm water in a bowl for 48 hours before sowing.

Fill the pots with soil-less mix, smooth over, tap down and water in well. Place 1 seed sideways in a pot and press into the compost, leaving the top edge exposed. Label with the plant name and date. Place the pot in a warm place in partial shade, at an optimum temperature of 54°F in shade. Do not allow the mix to dry out, but be careful not to overwater. Germination takes 3–9 months.

Pot up as soon as the seedlings are large enough to handle. Keep the young plants under protection and in partial shade for two years, then introduce slowly to strong sunlight. When potting up, use a coir or peat-based potting mix, mixed with extra coarse sand. Mix to a ratio of 3 parts mix + 1 part sand.

Alternatively, you can pregerminate cycad seeds. After soaking the seed for 48 hours (see above), put some damp (not wet) coir or sphagnum peat moss in a plastic bag and mix in the damp seeds. Seal the bag and write the name of the seed on the outside in waterproof pen. Then place this bag in light shade at an optimum temperature of 77°F. After 4 weeks, check regularly to see if there are any signs of sprouting. Pot up very carefully in order not to break the roots, using the potting mix described above. Place the pots in a warm humid position in partial shade. Grow on for a minimum of 4 years before planting out in a tropical climate. For those living in a cool climate, grow on as a conservatory plant using a potting mix made of 3 parts soil-less mix plus 1 part coarse grit.

Dioon *Cycadaceae*

A single genus of cycads that consists of subtropical evergreen shrubs. Plant in a well-drained soil, which is rich in humus, in a sunny position. In cool climates, this plant makes a very good houseplant, as it is quite happy with a few hours of good light per day. Ideal for a conservatory.

Dioon edule (Mexican cycad) Zones 9–10
Very large seed
Tropical evergreen shrub
Height up to 12 ft.
Female, pale-brown, oval, cone-like structure, which bears oval,
  nut-like seeds. A mature male and female cycad are needed
  to produce viable seeds
The leaves are light or bright green, blue or blue-green and
  semi-glossy. There are 50–150 leaves in a crown

In spring, soak the seed for 48 hours. Discard any floating seeds, then make a small nick in the end of the seed and sow immediately in deep pots, using standard soil-less seed mix, either coir or peat mixed with coarse grit. Mix to a ratio of 3 parts soil-less mix + 1 part coarse grit. Position the seed sideways, with only the top edge exposed. Cover with coarse grit, leaving the seed exposed, then place under protection at 70°F in shade. Germination takes 6–18 months.

Grow on as a container plant for 3 years, before planting out in warm climates. Alternatively, in cool climates, grow on as a container plant using a standard soil-less seed mix, either peat or peat substitute, mixed with coarse grit for extra aeration. Mix to a ratio of 3 parts soil-less mix + 1 part coarse grit.

The seed of this plant is often powdered and used as a kind of arrowroot.

Livistonia *Palmaceae*

A genus of, in the majority of cases, tall, tropical evergreen palms. Plant in light sandy loam in a sunny position. In cool climates, grow as a container plant.

Livistonia rountidifolia (Fan Palm) zone 10
Large seed
Tropical evergreen tree
Height up to 50 ft.
Greenish flowers in autumn, followed by oblong, smooth,
  blue/brown fruit in spring
Nearly round, dark-green, palm-shaped leaves deeply divided
  into 60–90 leaflets with slender points

In spring, remove the flesh carefully from the seed (see warning below), then soak the seed for 24 hours. Sow immediately into pots using standard soil-less seed mix, either coir or peat, mixed with $1/16$–$1/8$ in. fine grit. Mix to a ratio of 3 parts soil-less mix + 1 part fine grit. Cover with fine grit, then place under protection at 73°F. Germination takes 2–3 months.

Grow on as a container plant for 3 years, before planting out in warm climates. Alternatively, in cool climates, grow on as a container plant using a standard soil-less seed mix, either peat or peat substitute, mixed with extra fine potting bark for extra aeration. Mix to a ratio of 3 parts soil-less mix + 1 fine bark.

WARNING—Avoid contact with the fruit produced by this palm. It contains oxalic acid which is toxic when ingested, and contact with skin could result in severe chemical burns.

Phoenix *Palmaceae*

A genus of tropical and subtropical evergreen palms. Plant in a sandy soil in a sunny position. In cool climates, grow as a container plant.

Phoenix dactylifera (Date Palm) Zone 10
Large seed
Tropical evergreen tree
Height up to 100 ft. Rarely grows over 30 ft. in cool climates
The flowers, which are small, leathery and yellow, are followed by edible, yellow-orange to red fruit that ripen to dark brown.
Flowers in winter
Greenish or bluish-gray, pinnate leaves form a bushy canopy. The leaves are composed of long leaflets, which are arranged in v-shaped ranks and run the length of the leaf stem
Female plants will produce dates if a male tree is nearby

In spring, remove the flesh carefully from the seed, then soak the seed for 24 hours and sow immediately in pots, using standard soil-less seed mix, either coir or peat mixed with coarse sand to a ratio of 3 parts soil-less mix + 1 part coarse sand. Cover with coarse sand then place under protection at 68°F. Germination takes 1–2 months.

In warm climates, protect young plants from the sun for at least 2 years before planting out. Alternatively, in cool climates, grow on as a container plant using a standard soil-less seed mix, either peat or peat substitute, mixed with extra coarse sand for extra aeration. Mix to a ratio of 3 parts soil-less mix + 1 part coarse sand

---

Recipe for sowing, planting, and growing Phoenix dactylifera
Ingredients
2 seeds per pot
Small bowl
Hot, not boiling, water
1 x 4 in. pot
Standard soil-less seed mix, either coir or peat, mixed with coarse sand to a ratio of 3 parts soil-less mix + 1 part coarse sand
Waterproof pen
White plastic plant label
OR
1 plastic bag
Coir or sphagnum peat moss
Somewhere constantly warm

Method As soon as the fruits ripen, and change color from orange to brown, collect the seed, remove any pulp from around the seed and sow immediately. If you are using purchased, rather than fresh, seed, soak the seed in warm water in a bowl for 48 hours before sowing.

Fill the pots with soil-less mix, smooth over, tap down and water in well. Place 2 seeds in each pot, spaced equally on the surface of the mix. Cover the seeds with mix, to the same depth as the seed, and label with the plant name and date. Place the pot in a warm, light place out of direct sunlight at an optimum temperature of 68°F).

Do not allow the compost to dry out, but be careful not to overwater. Germination takes 1–2 months.

Pot up as soon as the seedlings are large enough to handle. Keep the young plants under protection and in partial shade for two years, then slowly introduce to strong sunlight. When potting up, use a coir or peat-based potting mix, mixed with extra coarse sand. Mix to a ratio of 3 parts soil-less mix, 1 part sand.

Alternatively, you can pregerminate palm seeds. After removing the pulp or soaking, put the seed with some damp (not wet) coir or sphagnum peat moss in a plastic bag, seal the bag and write the name of the seed on the outside in waterproof pen. Then place this bag somewhere where it is constantly warm, under a greenhouse bench for example.

After 4 weeks, check regularly to see if there are any signs of sprouting. Pot up, being very careful not to break the roots, using the potting mix described above.

Place the pots in a warm humid position in partial shade, and grow on for 2 years in partial shade.

Roystone *Palmaceae*

A genus of tropical evergreen palms. Plant in a sandy acid soil in a sunny position. These palms are one of world's most beautiful. They grace palaces and government buildings throughout the tropical world.

Roystone rigia (Royal Palm) Zones 9–10
Large seed
Tropical evergreen tree
Height up to 80 ft.
Creamy white, small flowers, followed by oval blue brown fruit
Flowers in winter
Terminal crowns of large, green, fan-shaped, pinnate leaves

In spring, remove the flesh carefully from the seed, then soak the seed for 24 hours and sow immediately into pots, using standard soil-less seed mix, either coir or peat, mixed with coarse sand to a ratio of 3 parts soil-less mix + 1 part sand. Cover with coarse sand, then place under protection at 68°F. Germination takes 2–3 months.

Grow on as a container plant for 2 years, before planting out in warm climates. Alternatively, in cool climates, grow on as a container plant using a standard soil-less seed mix, either peat or peat substitute, mixed with extra fine potting bark for extra aeration. Mix to a ratio of 3 parts soil-less mix + 1 part fine bark.

perennials

Perennials are wonderful plants for the majority of gardens as they team up so well with with trees, shrubs, and roses. Many species make attractive ground cover and, when grown in grand swathes, they look stunning in a bed. They provide an economical means of creating a display of color, texture, form, and structure. It is wonderful to see old favorites among the herbaceous perennials return after a hard winter, those first spring shoots are inspiring after the cold, dank, dark nights.

Many perennials will flower in their second season. It is usually worth sowing fresh seed for a more reliable germination rate, even though many can be stored for use in later years. It is also worth harvesting a few seeds of your favorite plant just in case you lose it in the winter.

Silene dioica seedheads, see page 168

Acanthus *Acanthaceae*

A genus of hardy perennials and semievergreens. Plant in a well-drained soil in a sunny situation, will adapt to partial shade.

Acanthus mollis (Bear's breeches) Zones 8–10
Large seeds: 14 per ounce
Hardy perennial, semievergreen
Height 6 ft.
Purple and white flowers on tall spikes
Flowers in high summer
Long, oval, deeply cut, bright green leaves

Sow in spring in pots, using standard soil-less seed mix, either peat or peat substitute. Cover with perlite or vermiculite, place under protection at 65°F. Germination takes 10-20 days. Protect young plants for the first winter. Flowers in its second or third year.

Achillea *Asteraceae*

A genus of hardy perennials and semievergreens, grown for their attractive flowers, which dry well. Plant in a well-drained soil in a sunny situation, although it will adapt to most soils and to partial shade.

Achillea filipendulina "Cloth of Gold" Zones 3–9
Small seeds: 17,360 per ounce
Hardy perennial
Height 5 ft.
Clusters of brilliant yellow flowers
Flowers in summer
Serrated, fern-like green leaves

Achillea millefolium "Cerise Queen" Zones 3–9
Small seeds: 14,000 per ounce
Hardy perennial
Height 2 ft.
Clusters of dark cherry-red flowers
Flowers in summer
Feathery, dark green leaves

Achillea ptarmica (Sneezewort) Zones 3–9
Small seeds: 12,040 per ounce
Hardy perennial
Height 2 ft.
Clusters of white flowers
Flowers in summer
Feathery, dark green leaves

Sow fresh seed in autumn in pots or cell packs, using standard soil-less seed mix, either peat or peat substitute. Cover lightly with soil-less mix and place in a cold frame. Germination takes 4-6 weeks. Overwinter young plants in the cold frame. Can flower in its first year.
OR
Sow in spring in pots or cell packs, using standard soil-less seed mix, either peat or peat substitute. Cover with perlite or vermiculite, place under protection at 68°F. Germination takes 10-15 days. Can flower in its first year.

Recipe for sowing, planting, and growing Achillea
Achillea is a lovely plant, which has pretty flowers that grow in clusters forming attractive flat heads in all shades of color. As the plants are available in all sizes they will suit most gardens from the alpine to the stately bed.

Ingredients
20 seeds per cell pack or 30 seeds per pot (do not sow thickly)
1 white 3 x 5 card, folded in half
1 flat with cell packs
OR
1 x 4 in. pot
Standard soil-less seed mix, either peat or peat substitute
Fine-grade perlite (wetted) or vermiculite
White plastic plant label

Method In autumn fill the cell packs or pot with soil-less mix, smooth over, tap down and water in well. As the seeds are fine, put a very small amount of seed into the crease of a 3 x 5 folded card and tap the card gently to sow. This will not only allow you to see what you are sowing, it will also enable you to sow thinly on the surface of the soil-less mix. Press the seeds gently into the mix with the palm of the hand. Cover with a very fine layer of mix. Label with the plant name and date. Place the flat or pot in a cold frame. Keep watering to the absolute minimum until germination has taken place. Germination takes 4–6 weeks. Overwinter the young seedlings undisturbed in the cold frame. Pot up or plant out in the spring, when there is no threat of frost, after a period of hardening off. Flowering will take place the following season.
OR
Sow the seeds in spring following the instructions above, except this time cover the seeds with the wetted fine-grade perlite or vermiculite and place the container in a warm light place out of direct sunlight and at an optimum temperature of 68°F. Keep watering to a minimum until germination has taken place. Germination takes 10–15 days with warmth. Prick out when the seedlings are large enough to handle or, if you are using cell packs, plant directly in a container or the garden after a period of hardening off and when there is no threat of frost. Many plants will flower in the first summer.

Aconitum *Ranunculaceae*

A genus of hardy perennials with hooded flowers. Plant in fertile, well-drained soil in a partially shaded situation.

Aconitum napellus (Monkshood, Wolf's Bane) Zones 3–8
ALL PARTS OF THIS PLANT ARE POISONOUS TO HUMANS
Medium seeds: 980 per ounce
Hardy perennial
Height 5 ft.
Tall spires of hooded, indigo blue flowers
Flowers in late summer
Deeply cut midgreen leaves

Sow in autumn in pots or cell packs, using standard loam-based seed mix. Cover lightly with soil mix, place in a cold frame. Germination takes 6 months and can be erratic. Flowers in its second year.

Aconitum napellus

OR
Sow in autumn in pots or cell packs, using standard soil-less seed mix, either peat or peat substitute. Cover with perlite or vermiculite. Place in the refrigerator for 8-12 weeks at 35–39°F. Remove and place under protection at 60°F. Germination takes 4–6 weeks after removing from the refrigerator. Flowers in its second year.

Agapanthus *Alliaceae*
A genus of perennials, some of which are evergreen, grown for their attractive blue flowers. Plant in a moist, well-draining soil in a sunny situation.

Agapanthus africanus (African lily) Zones 8–10
Medium seeds: 392 per ounce
Half-hardy evergreen perennial
Height 3 ft.

Umbels of deep blue flowers on upright stems
Flowers in summer
Broad dark green leaves

Agapanthus campanulatus
Medium seeds: 392 per ounce
Hardy perennial
Height up to 4 ft.
Rounded umbels of blue flowers on long stems
Flowers in summer
Narrow green leaves

Collected seeds may not come true to type, but they can yield some interesting variations. Sow in spring in pots or cell packs, using standard soil-less seed mix, either peat or peat substitute. Cover with perlite or vermiculite and place under protection at 61°F. Germination takes 18–24 days. Overwinter seedlings in cold frame and protect from frost. Flowers in its third year.

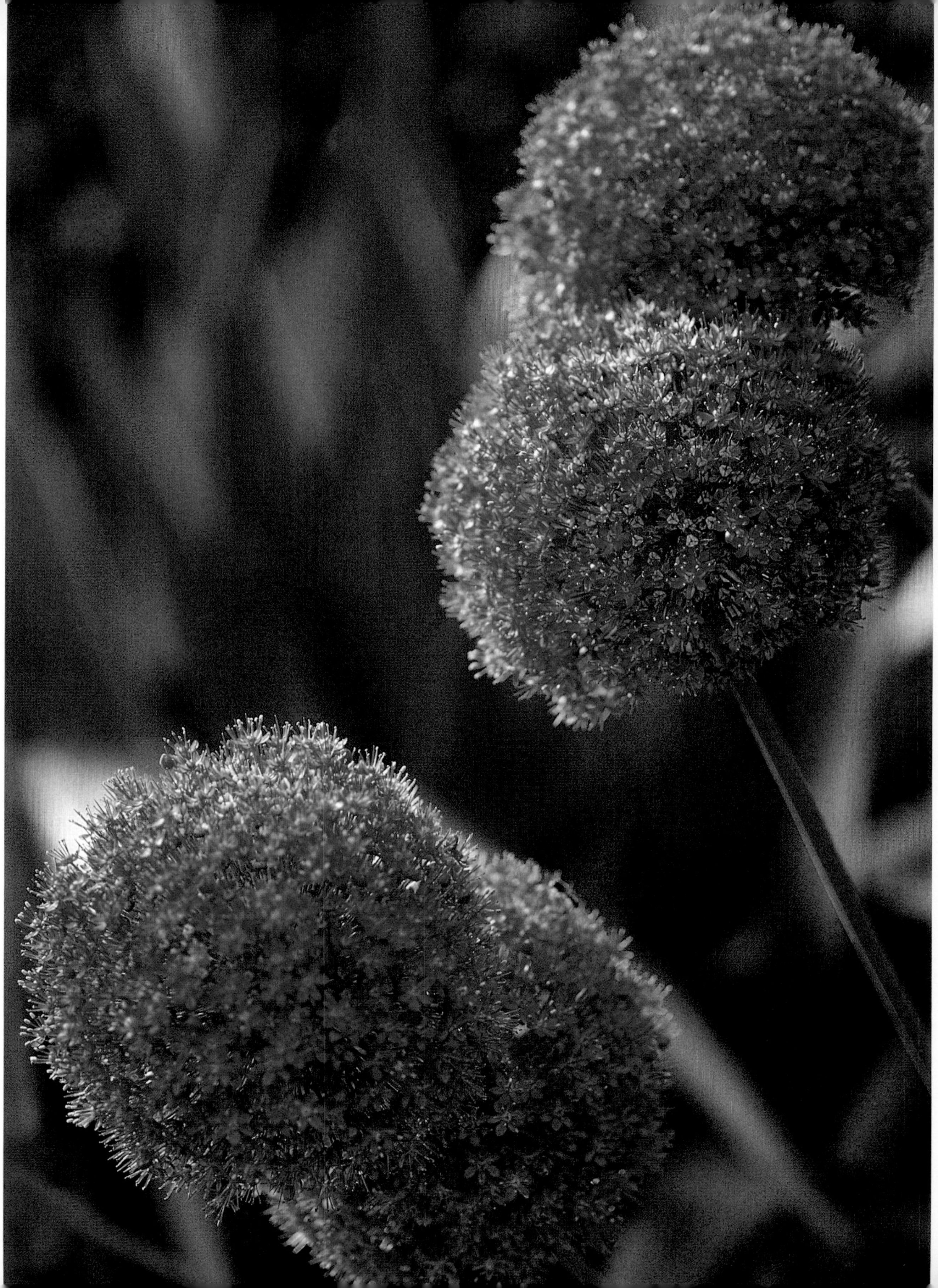

Allium giganteum

Agastache *Lamiaceae*

A genus of short-lived, summer-flowering perennials. Plant in a well-drained soil in a sunny situation.

Agastache foeniculum "Alba" (White anise hyssop) Zones 6–9
Small seeds: 8,400 per ounce
Hardy perennial
Height 2 ft.
Spikes of tubular white flowers
Flowers in summer
Oval, midgreen, aromatic, toothed leaves

Agastache urticifolia "Liquorice Blue" (Mexican Giant Hyssop)
Zones 7–9
Small seeds: 7,000 pr ounce
Hardy perennial grown as an annual
Height 2 ft.
Spikes of attractive ,blue mauve, tubular flowers
Flowers in summer
Oval, midgreen, toothed leaves

Sow in spring in pots or cell packs, using standard soil-less seed mix, either peat or peat substitute. Cover with perlite or vermiculite, place under protection at 68°F. Germination takes 7-10 days. Flowers in its first year.

Alchemilla *Rosaceae*

A genus of hardy perennials that have greenish-yellow flowers in summer and interesting leaves. Plant in sun or partial shade in most soils with the exception of bogs. Most species self-seed.

Alchemilla mollis (Lady's mantle) Zones 4–7
Small seeds: 8,120 per ounce
Hardy perennial
Height 22 in.
Greeny-yellow flowers in conspicuous outer calyces
Flowers in early summer
Rounded, pale green leaves with crinkled edges

Alchemilla vulgaris Zones 4–7
Small seeds: 4,480 per ounce
Hardy perennial
Height 12 in.
Greeny-yellow flowers in conspicuous, outer calyces
Flowers in early summer
Rounded pale green leaves, larger than *A. mollis*

Sow fresh seed in autumn in pots or cell packs, using standard loam-based seed mix. Cover lightly with mix, place in a cold frame. Germination takes 4-6 weeks and can be a bit erratic. Might flower in its first year.

Allium *Alliaceae*

This large genus of perennials, some of which are edible, can come in different forms—bulbs, rhizomes or fibrous rootstock. Nearly all have onion-smelling leaves, and most have small flowers which are clustered together in spherical (or similar) shapes. They are all hardy and the majority requires a sunny site with a well-drained, rich soil.

Allium giganteum Zones 4–8
Small seeds
Hardy perennial
Height up to 6 ft.
A dense, round umbel of 50 or more star-shaped purple flowers
Flowers in second summer
Wide, long, midgreen leaves

Sow fresh seed in autumn in pots or cell packs, using standard loam-based seed mix, mixed with coarse horticultural sand. Mix to a ratio of 1 part soil mix + 1 part sand. Cover lightly with mix, and place outside, exposed to all the weathers (see "Breaking Seed Dormancy", page 233, for more information). Germination takes place following spring. Might flower in its first year. In autumn these seed heads dry well and look great in arrangements.

Alstroemeria *Alstroemeriaceae*

A genus of summer-flowering tuberous perennials. Plant in a well-drained soil in a sunny, sheltered situation.

Alstroemeria ligtu hybrids Zones 7–9
Medium seeds: 168 per ounce
Hardy perennial
Height 2 1/2 ft.
Pink, rose and salmon lily-like flowers
Flowers from June until September
midgreen, lance-shaped leaves

Sow in autumn in pots or cell packs, using standard soil-less seed mix, either peat or peat substitute. Cover with perlite or vermiculite. Place under protection at 68°F for 3 weeks, then place containers in a cold frame or cold greenhouse at 40°F for a further 3 weeks. After this period of chilling, move back under protection at 68°F. Germination should occur within a further 4 weeks, however it can be erratic.

Althaea *Malvaceae*

A genus of annuals and perennials. Plant in a moist, fertile soil in a sunny situation.

Althaea officinalis (Marsh mallow) Zones 3–9
Medium seed: 1,120 per ounce
Hardy perennial
Height 2 3/4 ft.
Pink or white flowers
Flowers in late summer to early autumn
Gray green, tear-shaped leaves, covered with soft hair

Sow fresh seed in autumn in pots or cell packs, using standard soil-less seed mix, either peat or peat substitute. Cover lightly with soil-less mix, and place in a cold frame. Germination takes 2-4 months and can be erratic.

Calamintha nepeta (Lesser calamint) Zones 5–9
Small seeds: 9,240 per ounce
Hardy perennial
Height 15 in.
Loose clusters of small, tubular, pale lilac-pink to white flowers
Flowers all summer
Oval, toothed, green gray, mint-scented leaves

Sow seed in autumn in pots or cell packs, using standard loam-based seed mix, mixed with coarse horticultural sand. Mix to a ratio of 1 part soil mix + 1 part sand. Cover lightly with coarse horticultural sand and place in a cold frame. Germination takes 5–6 months.

These seeds can be tricky—the most successful method to date is the one mentioned above, although full stratification (see page 233) might work better for those in warmer climates.

Campanula *Campanulaceae*

A genus of annuals, biennials and perennials. Plant in a moist but well-drained soil in partial shade

Campanula glomerata "Superba" (Clustered bellflower) Zones 3–8
Small seeds: 25,200 per ounce
Hardy perennial
Height 2$\frac{1}{2}$ ft.)
Clusters of dark violet flowers
Flowers in early summer
Oval, green leaves in basal rosettes

Sow seed in spring in pots or cell packs, using standard soil-less seed mix, either peat or peat substitute. Place in a cold frame. DO NOT COVER. Germination takes 2–3 weeks. Divide and replant established plants regularly.

Canna *Cannaceae*

A genus of rhizomatous perennials. Plant in a humus-rich moist soil in a sunny sheltered position. Often used in garden displays and lifted for the winter.

Cannas are commonly propagated by division, however seed is not difficult and it is very rewarding,

Canna x generalis (Canna lily) Zones 8–10
Large seeds: 14 per ounce
Half-hardy perennial
Height 3 ft.
Spikes of stunning, orchid-like flowers. Colors vary from crimson to orange and yellow
Flowers in summer
Broad, bronze/green, lance-shaped leaves

In early spring, soak fresh seeds in hot, not boiling, water for 12 hours before sowing. Remove and discard any floating seeds and sow remainder in pots. Use standard soil-less seed mix, either peat or peat substitute. Cover with perlite or vermiculite and place under protection at 70°F. Germination takes 3–4 weeks. Will flower in its first year.

***Recipe for sowing, planting, and growing* Canna x generalis**
This is structurally a stunning plant; the foliage looks striking and the flowers display a wonderful range of colors. Well worth growing as a container plant if you live in cooler climates—make sure the container has a good base, so that the plant will not fall over in high winds.

Ingredients
2 seeds per pot
Small bowl
Hot water
1 x 4 in. pot
Standard soil-less seed mix, either peat or peat substitute
Fine-grade perlite (wetted) or vermiculite
White plastic plant label

Method In spring, fill a bowl with hot, not boiling, water. Add the seeds to the liquid and soak for 12 hours. Fill the pots with soil-less mix, smooth over, tap down and water in well. Place 2 seeds, equally spaced, on the surface of the mix in each pot. Cover the seeds with fine-grade perlite (wetted) or vermiculite and label with the plant's name and the date. Place the pot in a warm, light place, out of direct sunlight, at an optimum temperature of 70°F. Keep watering to a minimum until germination has taken place, which takes 3–4 weeks. Pot up as soon as the seedlings are large enough to handle. Keep the young plants under protection until large enough to stand outside or incorporate in a summer bedding scheme. As tender perennials, these plants will suffer if planted out before all threat of frost has passed.

Catananche *Asteraceae*

A genus of hardy, daisy-like, summer-flowering perennials. Plant in a light, well-drained soil in a sunny position.

Catananche caerulea (Cupid's dart) Zones 4–9
Medium seeds: 840 per ounce
Hardy perennial
Height 2 ft.
Attractive, daisy like, blue mauve flowers
Flowers all summer
Grass-like midgreen leaves
Flowers dry well for winter displays

Sow seed in spring in pots or cell packs, using standard soil-less seed mix, either peat or peat substitute. Cover with vermiculite or perlite and place under protection at 68°F. Germination takes 4–10 days. Will flower in its first year, although there is a better show in the second year.

Centaurea *Asteraceae*

A genus of hardy annuals and perennials, which all have attractive flowers. Plant in a well-drained soil and a sunny position.

Centaurea macrocephala Zones 3–7
Medium seeds: 168 per ounce
Hardy perennial
Height 3 ft.

Sow seed in early spring in pots or cell packs, using standard soil-less seed mix, either peat or peat substitute. Cover with perlite or vermiculite and place under protection at 65 °F. Germination takes 2–3 weeks.

Dahlia *Asteraceae*

A genus of half-hardy summer- and autumn-flowering tuberous perennials. There are many varieties of hybrid now available, which can be grown from seed. The flowers come in stunning colors and all shapes and sizes. Plant in a well-drained soil in a sunny position.

Dahlia "Figaro Red" Zones 8–9
Medium seeds: 280 per ounce
Half-hardy perennial often grown as an annual
Height 16 in.
Very attractive crimson flowers
Flowers all summer
Divided, midgreen, oval leaves

Sow seed in early spring in pots or cell packs, using standard soil-less seed mix, either peat or peat substitute. Cover with perlite or vermiculite, place under protection at 65°F. Germination takes 7–14 days.

Delphinium *Ranunculaceae*

A genus of perennials and annuals grown for their beautiful flowers. Plant in a rich, fertile well-drained soil in a sunny position.

Delphinium elatum Zones 3–7
Medium seeds: 980 per ounce
Hardy perennial
Height 4¹/₂ ft.–7 ft.
Spikes of semidouble flowers, regularly spaced, colors from white through to sky blue, to purple, with contrasting centers
Flowers in early summer
Large palmate, divided leaves

Delphinium nudicaule Zones 6–8
Medium seeds: 2,380 per ounce
Hardy, short-lived perennial
Height 8 in.
Spikes of hooded red or occasionally yellow flowers
Flowers in summer
Midgreen, basal, deeply divided leaves
Will flower in first year

Sow fresh seed individually in pots or cell packs in autumn, using standard soil-less mix, either peat or peat substitute, mixed with extra silver or fine sand for additional aeration. Mix to a ratio of 3 parts soil-less mix + 1 part sand. Cover with perlite or vermiculite, place in a cold frame. Germination takes 10–14 days. Overwinter young plants in a cold frame. Do not worry when the plant dies back, it is meant to. Plants will reappear in spring with lots of new, lush growth. However, beware of slugs.

Dodecatheon meadia

Dicentra *Papaveraceae*

A genus of hardy perennials grown for their elegant flowers. Plant in a moist but well-drained soil in semishade.

Dicentra spectabilis AGM (Bleeding heart) Zones 3–9
Medium seeds: 616 per ounce
Hardy perennial
Height 2¹/₂ ft.
Beautiful stalks with pendulous heart-shaped, pinkish-red and white flowers
Flowers in late spring
Fern-like, deeply cut midgreen leaves

Sow fresh seeds in autumn in pots, using standard loam-based seed mix mixed with coarse horticultural sand. Mix to a ratio of 1 part soil mix + 1 part sand. Cover lightly with coarse horticultural sand and place outside, exposed to all weathers (see "Breaking Seed Dormancy", page 233, for more information). Germination takes place the following spring. Flowers in its second season.

Dodecatheon *Primulaceae*

A genus of hardy perennials; plant in a moist but well-drained soil in sun or partial shade.

Dodecatheon meadia AGM
Small seeds: 5,040 per ounce
Hardy perennial
Height 12 in.
Large umbels of rose/purple flowers, with reflex petals and yellow anthers. The nickname of this flower is "shooting stars" because, once the flowers are pollinated, they turn their faces to the sky
Flowers from late spring
Rosettes of broad, light green, lance-shaped leaves

Sow fresh seeds in autumn in pots or cell packs, using standard soil-less seed mix, either peat or peat substitute mixed with silver or fine sand for extra aeration. Mix to a ratio of 3 parts soil-less mix + 1 part sand. Cover with fine sand and place in a cold frame. Germination takes 1–6 months, but can be erratic. This plant prefers to be planted in partial shade.

Doronicum *Asteraceae*

A genus of hardy perennials grown for their attractive daisy-shaped flowers. Plant in a well-drained soil in sun or light shade.

Doronicum orientale "Magnificum" (Leopards bane) Zones 4–7
Medium seeds: 2,800 per ounce
Hardy perennial
Height 20 in.
Large, cheerful, daisy-shaped, yellow flowers
Flowers in spring
Bright green, heart-shaped leaves

Loose spikes of very fragrant white, mauve, or yellow flowers
Flowers all summer from second season
Narrow lance-shaped midgreen leaves

In spring, soak the seeds in hot, not boiling, water overnight. Remove and discard any floating seeds and sow the remainder in pots or cell packs, using standard soil-less seed mix, either peat or peat substitute. Cover lightly with perlite or vermiculite and place under protection at 70°F. Germination takes 3–4 weeks.

Gaillardia *Asteraceae*

A genus of hardy annuals and short-lived perennials. Plant in a well-drained soil in a sunny situation.

Gaillardia x *grandiflora* "Dazzler" AGM (Blanket flower) Zones 3–9
Medium seeds: 700 per ounce
Hardy perennial
Height 2 ft.
Large, daisy-like, red flowers that have yellow tips to the petals
Does need staking
Flowers from summer until early autumn
Divided, soft, midgreen leaves

Sow seeds in spring in pots or cell packs, using standard soil-less seed mix, either peat or peat substitute. Cover with perlite or vermiculite, place in a cold frame. Germination takes 2–3 weeks.

Galega *Papilionaceae*

A genus of hardy summer-flowering perennials. Plant in well-drained soil in a sunny position.

Galega officinalis (Goat's rue) Zones 3–7
Medium seeds: 336 per ounce
Hardy perennial
Height 3 ft.
Spikes of small, pea-like, mauve flowers
Flowers in summer of second season
Oblong, lance-shaped leaves with oval leaflets

Sow fresh seeds in autumn in pots or cell packs, using standard soil-less seed mix, either peat or peat substitute. Cover with perlite or vermiculite and place in a cold frame. Germination takes 2–3 weeks. Overwinter young plants in a cold frame.

Galium *Rubiaceae*

A genus of hardy perennials. Plant in any well-drained soil, in partial shade – ideal for deciduous woodland.

Galium verum (Lady's bedstraw) Zones 4–8
Small seeds: 4,200 per ounce
Hardy perennial
Height 6 in.–3 ft.
Panicles of sweetly scented, tiny yellow flowers
Flowers in summer of second season
Whorls of small, linear leaves

Sow fresh seeds in late summer in pots or cell packs, using standard loam-based seed mix, mixed with coarse horticultural sand. Mix to a ratio of 1 part soil mix + 1 part sand. Cover with coarse horticultural sand and place in a cold frame. Germination takes 2–3 months and can be erratic.

Gaura *Onagraceae*

A genus of hardy biennials and perennials. Plant in a light, well-drained soil in a sunny situation.

Gaura lindheimeri AGM Zones 5–7
Medium seeds: 168 per ounce
Hardy perennial
Height 4 ft.
Racemes of tubular, white flowers that have a pink hue
Flowers in its first year
Lance-shaped midgreen leaves

Sow seeds in spring in pots or cell packs, using standard soil-less seed mix, either peat or peat substitute. Cover with perlite or vermiculite and place in a cold frame. Germination takes 2–3 months. Overwinter young plants in a cold frame.

Gazania *Asteraceae*

A genus of half-hardy perennials which are often grown as annuals. Plant in a well-draining sandy soil in a sunny situation.

Gazania "Daybreak Red Stripe" Daybreak Series (Treasure flower) Zones 8–10
Medium seeds: 1,120 per ounce
Half-hardy perennial, often grown as an annual
Height 8 in.
Large, daisy-like flowers in a mixture of colors—bronze, yellow, orange, pink, and white
Flowers in summer
Lance-shaped midgreen leaves

Sow seeds in spring in pots or cell packs, using standard soil-less seed mix, either peat or peat substitute. Cover with perlite or vermiculite and place under protection at 70°F or 60°F. Germination takes 10–14 days or 14–21 days at the cooler temperature.

Geranium *Geraniaceae*

A genus of hardy perennials, some of which are semievergreen. Plant in most soils, with the exception of waterlogged sites. Likes a sunny situation, although some species prefer a bit of shade.

Geranium pratense "Splish-splash" (Meadow cranesbill) Zones 5–8
Medium seeds: 364 per ounce
Hardy perennial
Height 20 in.
Saucer-shaped, 5-petaled blue/mauve/white speckled flowers
Flowers in summer from the second season
Rounded, lobed, deeply divided, midgreen leaves

Sow seeds in summer in pots or cell packs, using standard soil-less seed mix, either peat or peat substitute. Cover with perlite or vermiculite and place under protection at 60°F. Germination takes 10–14 days. Overwinter young plants in a cold frame.

Grevillea *Proteaceae*

A genus of half-hardy evergreen shrubs and trees. In warm climates, plant in a well-drained acid soil in a sunny situation. In cooler climates, grow as a potted plant and keep watering to a minimum in winter.

Grevillea robusta AGM (Silky oak) Zones 9–10
Medium seeds: 238 per ounce
Half-hardy evergreen perennial, often grown as a houseplant
Height in natural habitat 100 ft.
Mature plants have bell-shaped, yellow or orange flowers, which grow on one-sided spikes
Flowers all summer
Fern-like, long leaves

Sow seeds in early spring in pots or cell packs, using standard soil-less seed mix, either peat or peat substitute. Cover with perlite or vermiculite and place under protection at 65°F. Germination takes 3–4 weeks.
If growing on as a container plant in cool climates, use a peat or peat substitute potting mix.

Helenium *Asteraceae*

A genus of hardy perennials grown for their daisy-like flowers. Plant in a well-drained soil in a sunny situation.

Helenium autumnale
Small seeds: 11,200 per ounce
Hardy perennial
Height 3 ft.
Large red and gold daisy-shaped flowers
Flowers in summer and early autumn
Lance-shaped, midgreen leaves

Sow seeds in spring in pots or cell packs, using standard soil-less seed mix, either peat or peat substitute. Cover with perlite or vermiculite. Place in a cold frame. Germination takes 3–4 months. Overwinter young plants in cold frame. Flowers in its second season.

Heliopsis *Asteraceae*

A genus of hardy perennials. Plant in any free-draining soil in a sunny situation.

Heliopsis helianthoides var. *scabra* "Sommersonne" Zones 4–9
Medium seeds: 672 per ounce
Hardy perennial
Height 4 ft.
Large, golden yellow flowers
Flowers in summer
Narrow, oval, coarsely toothed, midgreen leaves

Sow seeds in spring in pots or cell packs, using standard soil-less seed mix, either peat or peat substitute. Cover with perlite or vermiculite and place in a cold frame. Germination takes 3–4 weeks.

Helleborus *Ranunculaceae*

A genus of hardy perennials, some of which are evergreen. Plant in a moisture-retentive, well-drained soil in semishade.

Helleborus argutifolius (corsicus) Zones 6–8
Medium seeds: 350 per ounce
Hardy perennial
Height 2 ft.
Large panicles of yellowish-green, cup-shaped flowers
Flowers in early spring
Thick, leathery, spiny, three-lobed leaves

Helleborus niger AGM (Christmas rose) Zones 3–8
Medium seeds: 350 per ounce
Hardy evergreen perennial
Height 12 in.
Cup-shaped, white flowers with golden stamens
Flowers from late winter to early spring
Mid-green, divided, oval leaves

Sow fresh seeds in early autumn in pots, using standard loam-based seed mix, mixed with coarse horticultural sand. Mix to a ratio of 1 part soil mix + 1 part sand. Cover with coarse horticultural sand and place under protection at 70°F for 6–7 weeks. Then place outside, exposed to all weathers (see "Breaking Seed Dormancy", page 233, for more information). Germination takes 5–12 months, so do not give up. If there is no germination in the first year, start the process again in the following autumn, with warmth for the first 6–7 weeks, after which, place the container outside.

Recipe for sowing, planting, and growing **Helleborus niger**
This beautiful winter-flowering plant is a must for any garden, it shows one that spring is not far away.

Ingredients
5 seeds per cell or 8 seeds per pot
1 flat with cell packs
OR
1 x 4 in. pot
Standard loam-based seed mix mixed with coarse horticultural sand to a ratio of 1 part soil mix + 1 part sand
Extracoarse horticultural sand to cover the pot
White plastic plant label

Method In autumn, fill the tray or pot with soil mix, smooth over, tap down and water in well. Sow the fresh seeds thinly on the surface of the soil mix. Press gently into the mix with the palm of the hand. Cover the seed with coarse horticultural sand. Label with the plant name and date. Place the flat or pot in a warm light place, out of direct sunlight, at an optimum temperature of 70°F.
After 6–7 weeks, place the tray or pot outside, on a level surface, so it is exposed to all weathers, including frosts. Do not worry if you

Helleborus orientalis

live in a snowy area and the containers get immersed in snow, as melting snow will aid germination. If you live in an area that will not get a winter frost, it is a good idea to put the container, after its period of heat, in a plastic bag, seal it and place in a refrigerator for 3 weeks. Remove the container from the refrigerator, then remove the plastic bag and place the container outside.

Whichever method you use, germination is a bit erratic, taking anything from 5–12 months. Do not give up and discard your soil mix, it might just germinate next week. If there is no germination in the first year, start the process again in the following autumn, with initial warmth for 6–7 weeks, followed by a period outside.

Prick out when the seedlings are large enough to handle. If you are using cell packs, you can plant directly in the garden as soon as the soil is warm enough to dig over.

Hemerocallis *Hemerocallidaceae*

A genus of hardy perennials. Plant in a fertile, well-drained soil in a sunny situation. When growing daylilies from seed, it is worth noting that cultivars will not come true to type, but will still give an interesting display. Also species will only come true if grown in isolation from other daylilies to prevent cross pollination.

Hemerocallis x *hybrida* (Daylily) Zones 3–9
Medium seeds: 98 per ounce
Hardy perennial
Height 18 in.
The trumpet-shaped flowers come in a mixture of colors from
    pale orange, yellow, and white to deep red and lavender pink
Flowers in summer from its second year
Midgreen, arching, strap-shaped leaves

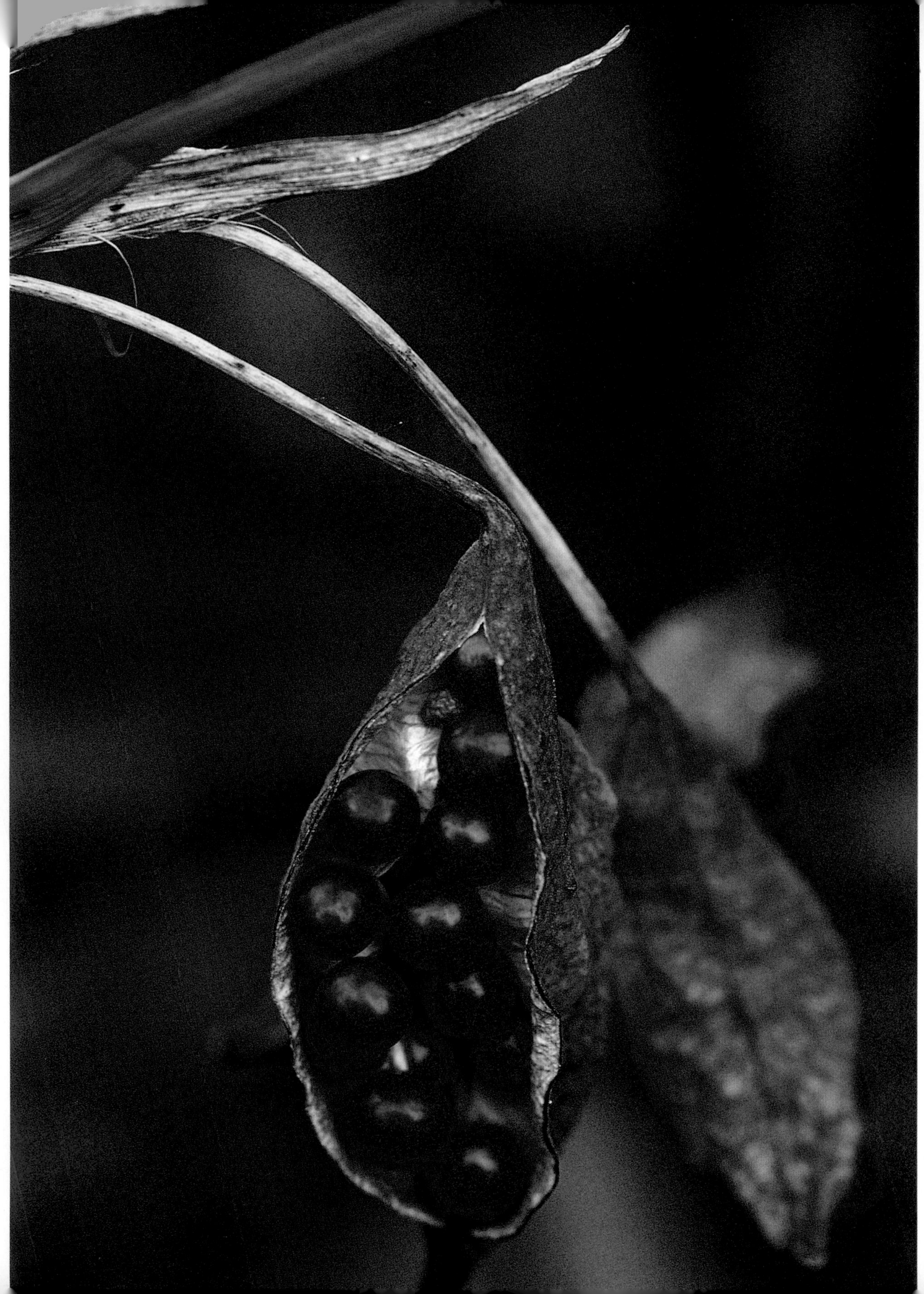

Iris foetidissima seeds

Sow seeds in early to midspring in pots or cell packs, using standard soil-less seed mix, either peat or peat substitute. Cover with perlite or vermiculite and place in a cold frame. Germination takes 3–4 weeks. A word of warning: protect young shoots from slugs.

Hosta *Hostaceae*

A genus of hardy perennials, grown for their attractive foliage. Plant in a rich, moist, well-drained neutral soil in partial to full shade.

Hosta sieboldiana Zones 3–8
Medium seeds: 812 per ounce
Hardy perennial
Height 36 in.
Racemes of trumpet-shaped, white flowers with a hint of lilac
Flowers in early summer
Large, heart-shaped, ribbed, puckered bluish-gray leaves

Sow seeds in early to midspring in pots or cell packs, using standard soil-less seed mix, either peat or peat substitute. Cover with perlite or vermiculite and place in a cold frame. Germination takes 1–3 months.

Incarvillea *Bignoniaceae*

A genus of hardy perennials. Plant in a well-drained soil in a sunny situation. It is worth protecting the crowns of this plant in winter with compost, leaf mold, or straw.

Incarvillea delavayi Zones 5–7
Medium seeds: 560 per ounce
Hardy perennial
Height up to 2¹/₄ ft.
Clusters of beautiful, bright rose-red, trumpet-shaped flowers, which are followed by attractive seed pods
Flowers from late spring until early summer
Deeply divided, midgreen leaves

Sow seeds in spring in pots or cell packs, using standard soil-less seed mix, either peat or peat substitute. Cover with perlite or vermiculite and place in a cold frame. Germination takes 3–4 weeks.

Iris *Iridaceae*

This very large genus is made up of rhizomatous or bulbous perennials, some of which are evergreen, most of which are hardy. There are irises suitable for the rock garden, woodland, waterside, alpine house, in fact most areas of the garden. For the ideal growing conditions, check with each species.

Iris foetidissima AGM (Galdwin, Stinking iris) Zones 6–9
Medium seeds: 56 per ounce
Hardy evergreen
Height 3¹/₂ ft.
Each branched stem can carry up to 9 yellow-tinged, dull purple, and occasionally pure yellow, flowers, which are followed by cylindrical seed pods with orange fruits

Flowers in summer
Spear-shaped, midgreen leaves
Plant in a bog or water garden

Iris setosa AGM (Bristle pointed iris) Zones 4–8
Medium seeds: 336 per ounce
Hardy evergreen
Height up to 3 ft.
Each branched stem can carry anything from 2–13 deep blue or purple blue flowers
Flowers in late spring to early summer
Spear-shaped, midgreen leaves

Scarify the fresh seed (see page 233) before sowing in early autumn in pots. Use standard loam-based seed mix mixed with coarse horticultural sand. Mix to a ratio of 1 part soil mix + 1 part sand. Cover with coarse horticultural sand and place under protection at 68°F for one month. Place outside, exposed to all weathers (see "Breaking Seed Dormancy", page 233, for more information). Germination takes 5–12 months, so do not give up. Young plants can winter outside happily. Flowers in the following season.

> **Recipe for sowing, planting, and growing Iris setosa**
> **Ingredients**
> 3 seeds per cell or 5 seeds per pot (sow 3 seeds per pot for *Iris foetidissima*)
> 1 sheet of fine sandpaper cut in half
> 1 flat with cell packs
> **OR**
> 1 x 4 in. pot
> Standard loam-based seed mix mixed with coarse horticultural sand to a ratio of 1 part soil mix + 1 part sand
> Extra sand for covering seeds
> White plastic plant label
>
> **Method** In autumn, fill the cell packs or pot with soil mix, smooth over, tap down and water in well. Select a small amount of seed, place it on half a sheet of fine sandpaper. With the other half of the sandpaper, either make a sandwich holding it between both hands and sliding back and forth to scratch the surface of the seed gently, or simply put the other half of the sandpaper on top of the seed and gently push the paper up and down which will have the same effect. Sow this scarified seed thinly on the surface of the soil mix. Press in gently with the palm of the hand and cover with coarse horticultural sand. Label with the plant name and date. Place the tray or pot outside, on a level surface, so it is exposed to all weathers, including frosts. Do not worry if you live in a snowy area and the containers get immersed in snow, for melting snow will aid germination. If you live in an area that will not get a winter frost, it is a good idea to put the scarified seed, mixed with a handful of damp sand, in a clearly marked plastic bag. Seal the bag and place it in a refrigerator for 3 weeks. Remove, and sow as described, then place outside. Germination will occur in the following spring and flowering the year after that.
>
> Prick out when the seedlings are large enough to handle. If you are using cell packs, you can plant directly in the garden as soon as the soil is warm enough to dig over.

Jasione *Campanulaceae*

A genus of hardy annuals, biennials, and perennials. Plant in a well-drained sandy soil in a sunny situation.

Jasione laevis (perennis) (**Sheep's bit**) Zones 6–8
Tiny seeds: 50,400 per ounce
Hardy perennial
Height 2–12 in.
Spherical, spiky, blue flower heads
Flowers in summer
Narrow, oblong gray green leaves

Sow fresh seeds in autumn in pots, using standard soil-less seed mix, either peat or peat substitute, mixed with extra silver or fine sand for additional aeration. Mix to a ratio of 3 parts soil-less mix + 1 part sand. As these are very fine seeds, mix with the finest sand or talcum powder for an even sowing. Do not cover. Water from the bottom or with a fine spray. Place in a cold frame. Germination takes 1–6 months. Overwinter young plants in cold frame. Flowers in its second year.
OR
Sow seeds in spring in pots or cell packs, using standard soil-less seed mix, either peat or peat substitute. As these are very fine seeds, mix with the finest sand or talcum powder for an even sowing. Do not cover. Water from the bottom or with a fine spray and place in a cold frame. Place under protection at 68°F. Germination takes 10–14 days.

Knautia *Dipsacaceae*

A genus of hardy annuals and perennials. Plant in a well-drained soil in a sunny situation.

Knautia arvensis (**Field scabious**) Zones 5–8
Medium seeds: 420 per ounce
Hardy perennial
Height up to 4 ft.
Blue lilac, pincushion-shaped flowers surrounded by petals
Flowers in summer
Deeply divided, midgreen leaves

Sow seeds in autumn in pots or cell packs, using standard loam-based seed mix, mixed with coarse horticultural sand. Mix to a ratio of 1 part soil mix + 1 part sand. Cover with coarse horticultural sand. Then place outside, exposed to all weathers (see "Breaking Seed Dormancy", page 233, for more information). Germination takes 4–6 months, but it can be erratic. Be patient and do not discard the soil mix.

Kniphofia *Asphodelaceae*

A genus of perennials, some of which are evergreen. Plant in a well-drained soil in a sunny position.

Kniphofia uvaria (**Red hot poker**) Zones 6–9
Medium seeds: 980 per ounce
Hardy perennial
Height 4 ft.
Tubular, bright red flowers in dense terminal racemes

Flowers in summer
Strap-shaped dark green leaves

Sow seeds in spring in pots or cell packs, using standard soil-less seed mix, either peat or peat substitute. Cover with perlite or vermiculite and place in a cold frame. Germination takes 2–3 weeks.
OR
Sow seeds in early spring (January–February) in pots or cell packs, using standard soil-less seed mix, either peat or peat substitute. Cover with perlite or vermiculite and place under protection at 65°F. Germination takes 14–21 days. Flowers in its first season.

Lantana *Verbenaceae*

A genus of tender evergreen perennials which needs a minimum temperature of 10°C (50°F). In warm climates, plant in light fertile well-drained soil in sun or partial shade. In cooler climates grow in containers and water sparingly in winter.

Lantana hybrida Crown Zones 8–10
Medium seeds: 140 per ounce
Half-hardy evergreen perennial
Height 3½ ft.
Tiny, tubular flowers that grow in dense clusters making a rounded head, in a range of pastel colors from cream through to pink, changing color as they mature
Flowers in summer
Wrinkled, dark green leaves
This is a good butterfly plant

Sow seeds in early spring in pots or cell packs, using standard soil-less seed mix, either peat or peat substitute. Cover with perlite or vermiculite and place under protection at 68°F. Germination takes 4–6 weeks. If growing on as a container plant, use a peat or peat substitute potting mix.

Leontodon *Asteraceae*

A genus of hardy annuals, biennials, and perennials. Plant in any soil, with the exception of waterlogged sites, in sun or partial shade.

Leontodon rigens (Microseris) Zones 6–9
Small seeds: 3,360 per ounce
Hardy perennial sometimes grown as an annual
Height 20 in.
Clusters of small, yellow, dandelion-like flowers
Flowers all summer
Rosettes of toothed, serrated, shiny green leaves

Sow seeds in spring in pots or cell packs, using standard soil-less seed mix, either peat or peat substitute. Cover with perlite or vermiculite, place in a cold frame or under protection at 68°F. Germination takes 2–4 weeks cold or 5–8 days with heat. Plants will flower in their first year.

Liatris *Asteraceae*

A genus of hardy summer-flowering perennials. Plant in a well-drained soil in a sunny position.

Liatris spicata (Blazing star) Zones 3–9
Medium seeds: 840 per ounce
Hardy perennial
Height 2 ft.
Spikes of purple rose flower heads on stiff stems
Flowers in summer
Grass-like, midgreen leaves

Sow seeds in spring in pots or cell packs, using standard soil-less seed mix, either peat or peat substitute. Cover with perlite or vermiculite and place in a cold frame or under protection at 65°F. Germination takes 3–4 weeks cold or 2–3 weeks with heat.

Lilium *Liliaceae*

A genus of, in the majority, hardy flowering bulbs. Plant in any well-drained soil in a sunny situation.

Lilium longiflorum AGM (Easter lily) Zones 6–9
Medium seeds
Hardy perennial
Height 1–3 ft.
1–6 fragrant, funnel-shaped, white flowers
Flowers in mid summer in second year
Lance-shaped leaves

Lilium regale (Regal Lily) Zones 3–8
Medium seeds: 1,260 per ounce
Hardy perennial
Height 20 in.–6 ft.
Can produce up to 25 fragrant funnel-shaped flowers, white on the inside, pinkish on the outside
Flowers in first or second summer
Lance-shaped leaves

Lilium regale

Growing plants from seed produced from a bulb-producing plant is a slow process but, with patience, you can produce plants in 1–2 years and flowers in 1–2 years. The major advantage of growing lilies this way is that they will be virus free.

Sow fresh (this is most important) seeds in autumn in pots, using standard loam-based seed mix mixed with coarse horticultural sand to a ratio of 1 part soil mix + 1 part sand. Cover with coarse horticultural sand. Place outside, exposed to all weathers (see "Breaking Seed Dormancy", page 233, for more information). Germination takes 4–6 months.

Leave undisturbed in the pots for the first season, allow the green growth to die back and stop watering. In the second year when the bulbs are dormant, repot them in fresh soil mix, the same mix as above, again winter them outside. In the following autumn, repeat the process, but this time the bulbs can be planted in the garden.

Lotus *Papilionaceae*

A genus of hardy to half-hardy perennials, some of which are evergreen. Plant in a well-drained soil in a sunny situation.

Lotus maritimus (Tetragonolobus) Zones 6–9
Medium seeds: 560 per ounce
Hardy perennial
Height 10 in.
Light yellow flowers, like a small sweet pea, which grow in profusion
Flowers in summer
Blue green foliage

Sow seeds in spring in pots or cell packs, using standard soil-less seed mix, either peat or peat substitute. Cover with perlite or vermiculite and place in a cold frame. Germination takes 3–4 weeks, but can be a bit erratic, occurring spasmodically over a long period.

Lupinus *Papilionaceae*

A genus of hardy annuals, perennials, and semievergreen shrubs. Plant in a well-drained, preferably alkaline, soil in a sunny position. Lupins are known for cross-pollinating.

Lupinus arboreus AGM (Tree lupin) Zones 8–9
Medium seeds: 126 per ounce
Hardy semievergreen shrub
Height 4–6 ft.
Short spikes of fragrant yellow flowers
Flowers in early summer of second or third season
Pale green leaves, which are divided into leaflets

Sow fresh seeds in autumn in pots or cell packs, using standard soil-less seed mix, either peat or peat substitute. Cover with perlite or vermiculite and place in a cold frame. Germination takes 2–3 weeks. Overwinter young plants in a cold frame.

Lupinus "The Governor" Zones 4–7
Medium seeds: 126 per ounce
Hardy perennial

Height 3 ft.
Large blue with white flower spikes
Flowers in summer of second season
Palmate, midgreen leaves

In autumn, soak fresh seeds in hot (not boiling) water for 12 hours before sowing. Remove and discard any floating seeds and sow the remainder in pots, using standard soil-less seed mix, either peat or peat substitute mixed with 1/8 in. grit. Mix to a ratio of 3 parts soil-less mix + 1 part fine grit. Cover lightly with perlite or vermiculite, place in a cold frame. Germination takes 2–5 weeks. Overwinter young plants in a cold frame. Do not worry when they die back, they will return in the spring. Protect young shoots from slugs.

I exhibit alongside the Woodfields Brothers at the Chelsea Flower Show, where they put on stunning displays of lupins, showing them to perfection. I am grateful to them for giving me these growing instructions.

Lysimachia *Primulaceae*

A genus of annuals and perennials. Plant in a moist but well-drained soil in sun or partial shade.

Lysimachia punctata (Garden loosestrife) Zones 4–8
Small seeds: 8,400 per ounce
Hardy perennial
Height 2½ ft.
Spikes of bright yellow, cup-shaped flowers
Flowers in summer
Broad, midgreen leaves

Sow seeds in spring in pots or cell packs, using standard soil-less seed mix, either peat or peat substitute. Cover with perlite or vermiculite and place in a cold frame. Germination takes 2–4 weeks.

Lythrum *Lythraceae*

A genus of hardy perennials. Plant in a moist or wet soil, in sun or partial shade – lovely around ponds.

Lythrum salicaria (Purple loosestrife) Zones 3–9
Tiny seeds: 36,400 per ounce
Hardy perennial
Height 2–4 ft.
Whorled spikes of bright pink/purple flowers
Flowers in summer
Willow-like leaves
Considered a noxious weed in some states

Sow fresh seeds in autumn in pots, using standard soil-less seed mix, either peat or peat substitute mixed with extra silver or fine sand for additonal aeration to a ratio of 3 parts soil mix + 1 part sand. Mix these very fine seeds with the finest sand or talcum powder for an even sowing. Do not cover. Water from the bottom or with a fine spray, place in a cold frame. Germination takes 5–7 months but can be erratic.

Recipe for sowing, planting, and growing Lythrum salicaria

This handsome and graceful plant looks lovely growing beside a stream or at the edge of a pond.

Ingredients
10 seeds per cell or 15 seeds per pot (or as near as you can manage)
Talcum powder or fine plain white flour
1 white 3 x 5 card, folded in half
1 flat with cell packs
OR
1 x 4 in. pot
Standard soil-less seed mix, either peat or peat substitute mixed with
    silver or fine sand for extra aeration. Mix to a ratio of 3 parts
    soil-less mix + 1 part sand
White plastic plant label

Method In autumn, fill the tray or pot with soil-less mix, smooth over, tap down and water in well. As the seeds are very fine, it is a good idea to mix them with talcum powder or extrafine white flour, which will make the seeds show up. Put a very small amount of this seed mix into the crease of the folded 3 x 5 card, gently tap the card to sow the seed thinly on the surface of the mix. Do not cover, label with the plant name and date. Place the flat or pot outside, on a level surface, so it is exposed to all weathers, including frosts. Do not worry if you live in a snowy area and the containers get immersed in snow, for melting snow will aid germination. If you live in an area that will not get a winter frost, it is a good idea to put the seed and flour mix and a handful of damp sand in a plastic bag. Seal and label the bag and place it in a refrigerator for 3 weeks. Remove, and sow as above, then place outside. Germination will occur in the following spring and flowering the year after that.

Prick out the seedlings when they are large enough to handle, harden them off before planting out. If you are using cell packs, you can plant directly in the garden when there is no further threat of frost.

Malva *Malvaceae*

A genus of hardy annuals, biennials, and short-lived perennials. Plant in a fertile, well-drained soil in a sunny position.

Malva sylvestris "Zebrina" Zones 3–8
Medium seeds: 644 per ounce
Hardy perennial
Height 4 ft.
Striking, bicolored flowers—pale pink/lilac, boldly marked with
    deep purple stripes
Flowers in summer
Divided, midgreen leaves

Sow in spring or late summer in pots or cell packs, using standard soil-less seed mix, either peat or peat substitute. Cover with perlite or vermiculite and place under protection at 68°F. Germination takes 6–10 days. Germination can be irregular.

Meconopsis *Papaveraceae*

A genus of hardy perennials, some of which are monocarpic (meaning that they die after flowering). Plant in a humus-rich, free-draining, neutral to acid soil in partial to fully shady areas. Succeeds only in the Pacific Northwest.

Meconopsis betonicifolia (syn. *baileyi*) Blue Himalayan poppy
Zones 6–8
Small seeds: 8,400 per ounce
Hardy perennial
Height 3–4 ft.
Beautiful, sky-blue, saucer-shaped flowers with yellow stamens
Flowers in summer
Rosettes of oblong, midgreen leaves at the base of the plant

Sow seeds in autumn in pots or cell packs, using standard loam-based seed mix, mixed with coarse horticultural sand. Mix to a ratio of 1 part soil mix + 1 part sand. Cover with coarse horticultural sand. Place outside, exposed to all weathers (see "Breaking Seed Dormancy", page 233, for more information). Germination takes 3–4 weeks. Overwinter young plants in a cold frame.

Penstemon *Scrophulariaceae*

A genus of annuals, perennials, and shrubs, most of which are evergreen or semievergreen. Plant in a fertile, well-drained soil in a sunny position.

Penstemon barbatus "Jingle Bells" Zones 3–8
Medium seeds: 2,520 per ounce
Hardy perennial
Height 5 ft.
Spikes of tubular, 2-lipped red flowers
Flowers in summer to early autumn
Rosettes of oblong, midgreen leaves

Sow seeds in spring in pots or cell packs, using standard soil-less seed mix, either peat or peat substitute. Cover with perlite or vermiculite and place in a cold frame. Germination takes 2–3 weeks. If no germination occurs, put under cover for one month at 70°F, then return to the cold frame.

Persicaria *Polygonaceae*

A genus of annuals and perennials, some of which are evergreen. Plant in a damp but well-drained soil in sun or partial shade.

Persicaria bistorta (syn. *Polygonum*) (Bistort, Snake Root)
Zones 3–8
Medium seeds: 392 per ounce
Hardy perennial
Height 2 1/2 ft.
Small spikes of soft pink flowers
Flowers in summer
Oval, midgreen leaves
This plant can be invasive

shrubs

Growing shrubs from seed is very satisfying. As there are very few commercial suppliers of seed, collecting your own is doubly worthwhile. It is great fun to collect the seed from a shrub that you admire in a friend's garden. Two years later you will be watching it flourish in your own garden.

Remember that most shrubs need to be grown on in containers for two years before planting out in their growing positions.

Shrub seed is very seldom sold by number, so in this chapter I have provided only an indication of the size. For more information on seed size see page 242.

Abutilon *Malvaceae*

A genus of annuals, perennials, evergreen, and semievergreen or deciduous shrubs. Plant in a fertile well-drained soil in sun or partial shade.

Abutilon megapotamicum AGM (Flowering maple) Zones 9–10
Medium seeds
Evergreen half-hardy shrub, often grown as a houseplant
Height 10 ft.
Pendant bell-shaped yellow and red flowers
Flowers in late spring to summer
Dark green, oval, slightly lobed leaves

Sow seeds in early spring in pots or cell packs, using standard soil-less mix, either peat or peat substitute. Cover with perlite or vermiculite, place under protection at 65°F. Germination takes 3–4 weeks. Flowers in two years.

Arctostaphylos *Ericaceae*

A genus of hardy to half-hardy evergreen trees and shrubs. Plant in a well-drained acid soil in a sunny situation.

Arctostaphylos uva-ursi (Bearberry) Zones 2–6
Medium seeds
Evergreen hardy shrub
Height 6 in.
Pink, urn-shaped flowers followed by brilliant red berries
Flowers in spring
Small oval, shiny, green leaves

In autumn, rub the berries in a rough cloth or use your thumbnail to remove the flesh from the seeds. Soak the seeds in hot water for 24 hours before sowing. Sow immediately, so that the seeds do not have a chance to dry out. Sow singly in pots or cell packs using peat or peat-based seed mix, mixed with horticultural sand. Mix to a ratio of 1 part mix + 1 part sand. Cover with coarse horticultural sand. Place outside exposed to all the weathers (see "Breaking Seed Dormancy", page 233, for more information). Germination takes 4-6 months, or even as long as 18, so be patient.

Aucuba *Aucubaceae*

A genus of evergreen shrubs. Plant in any soil, with the exception of waterlogged sites, in sun or shade.

Aucuba japonica Zones (6)7–9
Large seeds
Evergreen shrub
Height 8 ft.
Small purple/pink flowers, followed by bright red berries on female plants
Flowers in midspring
Lance-shaped dark green glossy leaves, often spotted with yellow

In autumn, rub the berries in a rough cloth or use your thumbnail to remove the flesh from the seeds. Sow the seed immediately so that they do not dry out. Sow individually in pots or cell packs, using standard loam-based seed mix, mixed with coarse horticultural sand. Mix to a ratio of 1 part mix + 1 part sand. Cover with coarse horticultural sand. Place outside, exposed to all weathers (see "Breaking Seed Dormancy", page 233, for more information). Germination takes 4–6 months, or even as long as 18, so be patient.

Berberis *Berberidaceae*

A genus of hardy, deciduous and evergreen, spiny shrubs. Plant in any soil, except for waterlogged sites, in sun or partial shade.

Berberis julianae Zones 6–9
Medium seeds
Evergreen hardy shrub
Height up to 10 ft.
Clusters of lightly scented yellow flowers followed by blue black berries in autumn
Flowers from late spring to early summer
Spine-toothed leaves, copper tinted when young

Chill the seed within the ripe fruit before sowing. Do this by mixing the fruit with sharp sand and then spreading this mixture in a seed flat. Leave outside for the winter to allow the fruit to decompose naturally. Alternatively, mix the fruit and sand, put the mixture in a plastic bag and place in the refrigerator for the winter.

Sow the clean seed in spring in pots or cell packs using standard loam-based seed mix. Cover with perlite or vermiculite and place in a cold frame. Germination takes 3–4 months.

***Recipe for sowing, planting, and growing* Berberis julianae**
This berberis with its fine, dense, spiky habit makes a very good hedge.

Ingredients
Seeds in berries
1 seed flat
Sharp sand
White plastic plant label
Waterproof pen
OR
Seeds in berries
1 plastic bag
Damp sand
Waterproof pen
1 plastic tie
1 sieve
5 seeds per pot
1 x 4 in. pot
Standard loam-based seed mix
Coarse grit
White plastic plant label

Method In autumn collect the ripe fruit, remove it from the stems, mix with sharp sand and spread the mixture on a seed flat. Label, and place the flat outside on a level surface, so that it is exposed to all weathers, including frosts. Do not worry if you live in a snowy area and the containers get immersed in snow, for melting snow will aid germination. If you live in a warm climate, you will need to put the sand and fruit mix into a plastic bag, label and seal it, and keep it in a refrigerator for the whole winter.

In spring, place the sand mixture, either from the seed flat or from the refrigerator, into a sieve and wash away the sand and the pulp, leaving only the clean seeds.

Fill some pots with standard loam-based seed mix to just below the ridge at the top of the pot. Sow the seeds on the surface of the mix, then cover with coarse grit to just below the rim. Label each pot.

Place the container in a cold frame. Germination should take a further 3–4 months, in good seasons it can be quicker. When the seedlings are large enough to handle, divide and repot singly in 4 in. pots and grow on in a cold frame. Plant out in the garden the following season.

Buddleia *Buddlejaceae*
A genus of hardy to half-hardy deciduous and evergreen shrubs and trees.

Buddleia davidii (Butterfly bush, Summer lilac) Zones 6–9
Minute seeds
Deciduous hardy shrub
Height up to 10 ft.
Dense clusters of lilac/purple flowers
Flowers in summer
Lance-shaped, gray green leaves

Sow fresh seeds in autumn in pots or cell packs using standard soilless mix, either peat or peat substitute. As these are very fine seeds, mix with the finest sand or talcum powder for an even sowing. Do not cover. Water from the bottom or with a fine spray. Place under protection at 68°F. Germination takes 10–20 days. Overwinter young plants under protection in a cold frame or unheated glass house. Will flower in its first year.

Callicarpa *Verbenaceae*
A genus of hardy deciduous shrubs. Plant in a fertile, well-drained soil in a sunny position.

Callicarpa bodinieri (Beautyberry) Zones 6–8
Small seeds
Deciduous hardy shrub
Height up to 10 ft.
Clusters of numerous star-shaped lilac flowers, followed by masses of small berries in shades of lilac, violet, and rich purple
Flowers in midsummer
Narrow leaves turning deep rose-purple in autumn

Chill the seed within the ripe fruit before sowing. Do this by mixing the fruit with sharp sand then spreading this mixture into a seed flat. Leave outside for the winter to allow the fruit to decompose naturally. Sow the clean seed in spring in pots or cell packs, using standard loam-based seed mix. Cover with perlite or vermiculite and place in a cold frame. Germination takes 3–4 months.

Camellia *Theaceae*
A genus of hardy to tender evergreen shrubs and trees. Plant in a well-drained neutral to acid soil in semishade and preferably sheltered from strong winds.

Camellia japonica (Common camellia) Zones (7)8–9
Small seeds
Evergreen shrub
Height up to 30 ft.
Double flowers in shades of white, pink, and red
Flowers in spring to early summer
Lance-shaped, dark green leaves

Sow the fresh seed as soon as the fruits split revealing the seed. Sow in autumn in pots or cell packs, using peat moss or peat-based seed mix. Cover with perlite or vermiculite, place in a cold frame. Germination takes 2–6 months. Keep young plants in a cold frame for their first year.

Carissa *Apocynaceae*
A genus of tender, evergreen shrubs. In warm climates plant out in a well-drained soil in partial shade. In cool climates grow as a container plant.

Carissa macrocarpa (*grandiflora*) (Natal plum)
Medium seeds
Evergreen tender shrub (minimum temperature 50°F
Fragrant, white jasmine-like flowers, followed by edible plum-like fruits
Flowers in summer
Dark green glossy leaves and thorny stems

Sow fresh seed in autumn in pots or cell packs, using standard soilless mix, either peat or peat substitute. Cover with perlite or vermiculite, place under protection at 65°F. Germination takes 14–21 days. Overwinter young plants in a frost-free environment above the minimum temperature.

Caryopteris *Verbenaceae*
A genus of hardy deciduous shrubs. Plant in a light, well-drained soil in a sunny situation.

Caryopteris incana Zones 7–9
Small seeds
Deciduous hardy shrub
Pretty, scented tubular blue/violet flowers
Flowers in summer
Lance-shaped, gray green leaves

Sow fresh seed in spring in pots or cell packs, using standard soilless mix, either peat or peat substitute. Cover with perlite or vermiculite. Place under protection at 65°F. Germination takes 14–21 days. Flowers in 2–3 years from germination.

Collect berries in late winter and extract the seeds (average 2 per berry).

Sow seed in pots or cell packs, using standard, loam-based seed mix, mixed with coarse horticultural sand to a ratio of 1 part soil mix + 1 part sand. Cover with coarse sand and place under protection at 65°F. for 4 weeks. Then place outside to expose to all weathers. (See "Breaking Seed Dormancy", page 233, for more information.) Germination can occur from spring until the following spring. Do not give up.

Cytisus *Papilionaceae*

A genus of hardy to half-hardy deciduous and evergreen shrubs. Plant in a light, fertile, well-drained soil in a sunny position.

Cytisus scoparius (Broom, Common broom) Zones 5–8
Medium seeds
Deciduous hardy shrub
Profusions of bright, yellow, pea-like flowers
Flowers in early summer
Small, narrow, dark green, divided leaves

In autumn soak the seeds in hot, not boiling, water overnight. Remove and discard any floating seeds and sow the remainder in pots or cell packs, using standard loam-based seed mix, mixed with coarse horticultural sand to a ratio of 1 part soil mix + 1 part sand. Cover with coarse horticultural sand and then place outside exposed to all the weathers. (See "Breaking Seed Dormancy" for more information.) Germination takes 4–6 months.

Daphne *Thymelaeaceae*

A genus of hardy evergreen and deciduous shrubs. Plant in a well-drained fertile soil in sun or partial shade.

Daphne mezereum Zones 4–8
Large seeds
Deciduous hardy shrub
Height up to 4 ft.
Clusters of scented, purple/pink blooms followed by red fruits
Flowers from late winter until early spring, fruits appearing in spring
Small, oval, gray green leaves

Sow fresh seed in spring, having removed the pulp with a rough cloth or thumbnail, in pots or cell packs, using standard loam-based seed mix, mixed with coarse horticultural sand. Mix to a ratio of 1 part soil mix + 1 part sand. (Do not allow the seeds to dry out.) Cover with coarse horticultural sand. Place in a cold frame. Germination takes 4–6 months. Overwinter young plants in a cold frame.

Elaeagnus *Elaeagnaceae*

A genus of hardy deciduous and evergreen shrubs or trees. Plant in fertile, well-drained soil in sun or partial shade.

Elaeagnus pungens Zones 6–9
Medium seeds
Evergreen hardy shrub
Height up to 10 ft.

Clusters of fragrant, cream white flowers, followed by red, edible fruit
Flowers in mid to late autumn
Lance-shaped, dark green glossy leaves

Sow fresh seed in spring, having removed the pulp with a rough cloth or thumbnail. Sow in pots or cell packs, using standard loam-based seed mix, mixed with coarse horticultural sand to a ratio of 1 part soil mix + 1 part sand. Cover with coarse horticultural sand. Place in a cold frame. Germination takes 1–2 months, but can take longer, so if necessary leave in container for a further season. Overwinter young plants in a cold frame.

Fatsia *Araliaceae*

A genus of one species, see below. Plant in a fertile, well-drained soil in sun or partial shade. In cold areas protect from strong winds.

Fatsia japonica Zones 8–10
Small seeds
Evergreen hardy shrub
Height up to 10 ft.
Dense clusters of small, white flowers, followed by black berries in winter
Flowers in midautumn
Large, lobed, glossy, dark green leaves

Sow fresh seed extracted from the ripe black fruit in early winter in pots or cell packs, using standard soil-less mix, either peat or peat substitute. Cover with perlite or vermiculite. Place under protection at 65°F. Germination takes 10–21 days. Grow seedlings on under protection and protect young plants from frost for two winters.

Forsythia *Oleaceae*

A genus of hardy deciduous shrubs, which usually have masses of yellow flowers that open before the leaves appear. Plant in a fertile, well-drained soil in a sunny position.

Forsythia suspensa
Small seeds
Deciduous hardy shrub
Height up to 6 ft.
Small, trumpet-shaped, bright yellow flowers before leaves appear
Flowers in early spring
Small midgreen leaves

Collect seeds in late summer. Sow fresh seed in early spring in pots or cell packs, using standard loam-based seed mix, mixed with coarse horticultural sand. Mix to a ratio of 1 part soil mix + 1 part sand. Cover with coarse horticultural sand. Place in a cold frame. Germination takes 1–2 months.

Fremontodendron *Steruliaceae*

A genus of evergreen or semievergreen shrubs grown for their attractive flowers. Plant in a well-drained, light soil in a sunny position. In cold areas plant against a warm wall.

Fremontodendron californica (Flannel bush) Zones 9–10
Medium seeds
Evergreen shrub
Height up to 20 ft.
Large, saucer-shaped bright yellow flowers
Flowers from late spring to midautumn
Lobed dark green leaves.

Sow fresh seed in spring in pots or cell packs, using standard soilless mix, either peat or peat substitute. Cover with perlite or vermiculite and place under protection at 65°F. Germination takes 4–5 weeks. Do not overwater young seedlings, as they are prone to "damping off".

Fuchsia *Onagracea*

A genus of hardy to tender deciduous or evergreen shrubs and trees. Plant in fertile, moist, but not overwet, soil in partial shade. This is a large family so there are exceptions to the general rules. It is always worth talking to a fuchsia specialist about individual varieties. In cool climates, tender plants should be grown as container plants.

Fuchsia magellanica (Lady's ear drops) Zones 6–9
Small seeds
Deciduous hardy shrub
Height up to 10 ft
Small flowers, with red tubes and long red sepals and purple petals, followed by black fruits
Flowers in summer
Midgreen oval pointed leaves

Fuchsia procumbens Zones 7–9
Small seeds
Deciduous hardy shrub
Height up to 4 in.
Masses of tiny, upright, yellow tubed flowers with purple sepals and bright blue pollen, followed by red berries
Flowers in summer
Dark green oval pointed leaves

Collect the seeds from the fleshy fruit in late winter and store in a sealed plastic bag placed in the refrigerator. Sow chilled seed in spring in pots or cell packs, using standard soil-less mix, either peat or peat substitute. Cover with perlite or vermiculite and place under protection at 68°F. Germination takes 3–4 weeks. Can flower in its first year and will certainly flower in its second.

Gardenia *Rubiaceae*

A genus of tender, evergreen shrubs and trees with lovely flowers. Plant in a humus-rich neutral to acid soil in partial shade. In cool climates grow as a container plant in peat-based soil mix.

Gardenia jasminoides (*G. florida*) Zones 8–10
Small seeds
Evergreen tender shrub (minimum temperature 60°F)
Height up to 5 ft.

Strongly scented, pure white flowers
Flowers from summer until winter
Oval, dark green glossy leaves

Collect seeds in winter and sow in the spring in pots or cell packs, using peat moss or a peat-based mix. Cover with perlite or vermiculite and place under protection at 68°F. Germination takes 1–3 weeks. Grow on with protection at all times. The plants will take up to 7 years to flower.

Hamamelis *Hamamelidaceae*

A genus of hardy deciduous shrubs. Plant in a well-drained, fertile, peaty, acid soil in sun or semishade.

Hamamelis virginiana (Witch hazel, Virginian witch hazel)
Zones 3–8
Small seeds
Deciduous hardy shrub
Height up to 12 ft.
Small, fragrant yellow flowers
Flowers from midwinter until spring
Oval midgreen leaves

In cool climates, collect seed capsules just as they turn brown in autumn. Put them in a paper bag, seal the bag, and put it in a warm place. When ripe and dry, the seed capsules will explode.

Sow fresh seed in spring in pots or cell packs, using standard soil-less mix, either peat or peat substitute, mixed with coarse horticultural sand to a ratio of 1 part soil-less mix + 1 part sand. Cover with coarse horticultural sand. Place outside exposed to all weathers (see "Breaking Seed Dormancy", page 233, for more information). Germination will occur the following spring. Overwinter young plants in a cold frame the following year.

Hibiscus *Malvaceae*

A genus of hardy to tender perennials, annuals, shrubs, and trees, either evergreen or deciduous. Plant in a humus-rich, well-drained soil in a sunny position.

Hibiscus mutabilis (Cotton rose, Confederate rose) Zones 8–10
Medium seeds
Evergreen tender shrub/tree (minimum temperature 40°F)
Height up to 15 ft.
Funnel-shaped, white or pink flowers that darken with age and occasionally can be double
Flowers from summer until early autumn
Rounded lobed leaves

Collect seed from large, dry capsules in autumn.

Sow fresh seed in spring in pots or cell packs, using standard soil-less mix, either peat or peat substitute. Cover with perlite or vermiculite place under protection at 68°F. Germination takes 1–3 weeks. Grow on under protection.

Hydrangea quercifolia

Hydrangea *Hydrangeaceae*

A genus of hardy, deciduous shrubs and climbers, which can also be evergreen.

Hydrangea quercifolia AGM (Oak-leaved hydrangea) zones 5–9
Small seeds
Deciduous hardy shrub
Height up to 6–8 ft.
Clusters of small white flowers
Flowers in summer until autumn
Deeply lobed, green leaves that, in autumn, turn red and purple

Collect seed from dried capsules in late autumn/early winter. Sow fresh seed in spring in pots or cell packs, using standard soil-less seed mix, either peat or peat substitute. Cover with perlite or vermiculite and place under protection at 50°F. Germination takes 1–3 weeks.

Ligustrum *Oleaceae*

A genus of hardy deciduous, semievergreen, and evergreen shrubs and trees. Plant in any well-drained soil in sun or partial shade.

Ligustrum ovalifolium (Oval-leaved privet) zones 5–9
Medium seeds
Hardy evergreen shrub
Height up to 10 ft.
Clusters of small, dull, white flowers, that are followed by round shiny black berries
Flowers in summer
Oval smooth darkish-green leaves

Collect seeds from ripe berries in autumn (1–4 seeds per berry). Mix the seeds with a small amount of damp vermiculite, put the mixture in a plastic bag. Seal, label and put in a refrigerator for 3 weeks.

Nerium subsp. *oleander*

Sow the chilled seeds in winter in pots or cell packs, using standard loam-based seed mix mixed with coarse horticultural sand. Mix to a ratio of 1 part seed mix + 1 part sand. Cover with coarse horticultural sand, then place outside exposed to all weathers (see "Breaking Seed Dormancy", page 233, for more information). Germination takes 4–6 months. Grow on in a cold frame for the next season.

Mahonia *Berberidaceae*

A genus of hardy to half-hardy evergreen shrubs, grown for their foliage. Plant in a fertile, well-drained soil—which does not dry out—in summer, in partial shade or even full shade.

Mahonia acanthifolia Zones 8–10
Medium seeds
Evergreen shrub
Height up to 12 ft.
Long chains of rich yellow flowers, followed by blue/black berries
Flowers from late autumn to early winter
Dark green, holly-like spiny leaflets

Collect the ripe fruits in early summer, remove the seeds and wash thoroughly in running water. Sow in pots or cell packs, using standard loam-based seed mix, mixed with coarse horticultural sand. Mix to a ratio of 1 part soil mix + 1 part sand. Cover with coarse horticultural sand. Place in a cold frame. Germination takes 1–2 months but can take longer, so leave in container for a further season. Overwinter young plants in a cold frame.

Mimosa *Mimosaceae*

A genus of tender shrubs (minimum temperature 50°F) grown for their flowers. Plant in a well-drained soil in a sunny situation. In cool climates grow as a container plant.

Mimosa pudica (Sensitive plant)
Medium seeds
Tender evergreen shrub (minimum temperature 50°F)
Height 12 in.
Pretty, fluffy, pink flowers that close when touched
Flowers from summer to autumn
Small pinnate foliage, which curls when touched

Collect the hard-coated seeds in late summer when ripe.

In spring, soak the seeds in hot, not boiling, water for 12 hours before sowing in pots or cell packs, using standard loam-based seed mix, mixed with coarse horticultural grit. Mix to a ratio of 1 part soil mix + 1 part 1/4 in. grit. Cover with sharp grit. Place the container under protection at a temperature of 65°F. Germination takes 3–4 weeks. Flowers during second season. Overwinter the young plants under protection, minimum temperature 50°F.

Myrtus *Myrtaceae*

A genus of hardy to half-hardy evergreen shrubs, grown for their attractive flowers and aromatic foliage. Plant in a well-drained soil in a sunny position.

Myrtus communis AGM (Myrtle) Zones 9–10
Medium seeds
Evergreen shrub
Small, white flowers, followed by edible, black, oval berries
Flowers in summer
Small, oval, aromatic dark green leaves

Sow fresh seed in spring, having removed the pulp with a rough cloth or thumbnail. Sow in pots or cell packs, using standard loam-based seed mix, mixed with coarse horticultural sand. Mix to a ratio of 1 part soil mix + 1 part sand. Cover with coarse horticultural sand and place under protection at 60°F. Germination takes 1–2 months, but can take longer.

Nerium *Apocynaceae*

A genus of tender shrubs (minimum temperature 50°F) grown for their flowers. Plant in well-drained soil in a sunny situation. In cool climates, grow as a container plant.

Nerium oleander Zones 9–10
Medium seeds
Tender evergreen shrub (minimum temperature 50°F)
Height up to 12 ft.
Clusters of pink, white, red, apricot flowers
Flowers from spring until autumn
Lance-shaped, leathery green leaves

Collect seeds from bean-like pods in autumn.

Sow in spring or late summer in pots or cell packs, using standard soil-less mix, either peat or peat substitute. Cover with perlite or vermiculite, place under protection at 65°F. Germination takes 2–3 weeks. Grow on under protection at 150°F.

Paeonia *Paeoniaceae*

A genus of hardy perennials and deciduous shrubs grown for their showy flowers and foliage. Plant in fertile, well-drained soil in a sunny position. Can tolerate partial shade.

Paeonia delavayi AGM (Tree peony) Zones 5–9
Large seeds
Hardy deciduous shrub
Height up to 6 ft.
Bowl-shaped, single, deep red flowers
Flowers in spring
The midgreen leaves are divided into pointed oval leaflets

This seed has a double dormancy, producing roots in its first year and leaves in its second. It needs two cold periods, with warmth in between.

Collect ripe, fresh seeds in early autumn.

Sow individually in pots, using standard soil-less seed mix, either peat or peat substitute mixed with coarse horticultural sand. Mix to a ratio of 1 part soil-less mix + 1 part sand. Cover with coarse grit, then place outside exposed to all weathers (see "Breaking Seed Dormancy",

Outside my children's primary school there is a horse-chestnut tree, *Aesculus hippocastanum*, which was planted by one of my neighbor's sons, now a grown man. Recently, he returned to the village with his family and took his own children to see the tree that he had planted at their age. It now towers over the school fence, providing shade in summer, a good supply of buckeyes in autumn, and giving endless pleasure to the children playing in the school grounds. To me, the most amazing feat of the seed is to produce a tree that can grow to over 100 ft. and live for more than a century, giving pleasure to many generations.

Being so large, trees are slower to grow than flower seeds, and need a little more care while growing on, before planting out in their permanent position. There are very few tree seeds commercially available, but they are not difficult to collect, and are all best sown fresh. If you have a refrigerator large enough, it is worth storing the seeds in damp vermiculite and sowing later in winter, so protecting your harvest from predators, such as mice and squirrels.

trees

Acacia *Mimosaceae*

A genus of evergreen or semievergreen deciduous trees and shrubs. Grown for their attractive flowers. Plant in a well-drained soil in a sunny situation.

Acacia longifolia (Sidney golden wattle) Zones 8–10
Medium seeds
Evergreen tree
Height up to 20 ft.
Pendant clusters of golden yellow flowers
Flowers in spring
Oblong, narrow, dark green phyllodes

Before sowing, rub the hard-coated seeds gently with sandpaper then soak for 24 hours in hot, not boiling, water. Sow the seeds in early spring in pots or cell packs, using standard loam-based seed mix. Cover with vermiculite or mix and place under protection at 60°F. Germination takes 3-4 weeks. Grow on as a container plant in soil-based potting mix for 2 years, then plant out in the garden. This plant will flower when it is 3-4 years old. It is best grown in large pots and overwintered with protection.

Acer *Aceraceae*

A genus of hardy evergreen or deciduous trees and shrubs that are renowned for their autumn colors. Plant in a fertile, well-drained soil in sun or partial shade.

Acer platanoides (Norway maple) Zones 3–7
Medium-sized winged seeds
Deciduous hardy tree
Height up to 80 ft.
Clusters of small yellow flowers before the leaves appear
Flowers in spring
Broad, lobed green leaves that turn yellow in autumn

Acer pseudoplatanus (Sycamore) Zones 4–7
Medium-sized winged seeds
Deciduous hardy tree
Height up to 115 ft.
Clusters of small, pendulous green/yellow flowers
Flowers appear with the leaves in spring
Broad, 5-lobed green leaves

It is important that this seed does not dry out, either sow fresh in autumn or the following spring. If sowing seeds in spring, soak them in hot, not boiling, water for 48 hours.

Sow the seeds in pots or cell packs, using standard loam-based seed mix. Cover with mix, then a thin layer of coarse grit, and place in a cold frame. Germination occurs when the temperature reaches a constant 50°F. It is worth noting that, in some years, the seed will not germinate until the second spring.

Pot up when seedlings are large enough to handle. Plant in its growing position 2 years after germination.

Aesculus *Hippocastanaceae*

A genus of hardy deciduous trees and shrubs, grown for their bold flowers and fruits. Plant in a fertile, well-drained soil in sun or partial shade.

Aesculus hippocastanum (Horse-chestnut, Buckeye) Zones 3–7
Large seeds
Deciduous hardy tree
Height up to 115 ft.
Spikes of white flowers, followed by spiny fruits which contain one or more shiny buckeyes
Flowers in early summer
Leaves have 5 to 7 large, thick, stalkless leaflets

Sow fresh buckeyes in autumn in individual pots, using standard loam-based seed mix. Cover with mix, and then a layer of coarse grit.

Put outside exposed to all weathers (see "Breaking Seed Dormancy", page 233 for more information). Germination will occur the following spring, when the temperature reaches a constant 50°F.

Alternatively, if you cannot sow immediately on day of collection, remove the husks from the buckeye and place it in a plastic bag filled with moist peat, seal and store at 37°F. Sow in winter, following directions above. Plant in growing position 18 months after germination.

***Recipe for sowing, planting, and growing* Aesculus hippocastanum**

The horse-chestnut tree is very easy to raise from seed and will convince anyone that growing plants from seed is a wonderful thing.

Ingredients
1 seed per pot
1 x 4 in. diameter pot
Standard loam-based seed mix
Coarse grit
White plastic plant label
OR
1 seed
1 plastic bag
Damp sand
Waterproof pen
1 wire twist tie

Method Collect the buckeyes in autumn, remove the outer green husk. Half fill the pot with soil mix, place the buckeye, shiny side up, on the surface of the mix, cover with the remaining soil mix to below the ridge of the pot. Cover the surface of the mix with grit to just below the rim of the pot. Water in well. Label the pot with the date and name. Place the pot outside on a level surface, exposed to all weathers, including frosts. Do not worry if you live in a snowy area and the containers get immersed in snow, melting snow will aid germination. If you live in an area that will not get a winter frost, it is a good idea to put the seed and a handful of damp sand into a plastic bag which is clearly marked. Seal the bag. Place in a refrigerator for 3 weeks. Remove and sow as instructed above, then place outside. Make sure the pot does not dry out, and guard against raids by small rodents. Germination should occur the following spring. Allow the seedling to become fully rooted in the container before planting out 18 months later.

Aesculus hippocastanum

Ailanthus *Simaroubaceae*

A genus of hardy deciduous trees. Plant in a deep, fertile, well-drained soil in sun or partial shade.

Ailanthus altissima AGM (Tree of heaven) Zones 4–8
Medium-sized winged seeds
Deciduous hardy tree
Height up to 80 ft.
Clusters of small green flowers that are followed by winged seeds. Male and female flowers usually grow on separate trees
Flowers in midsummer
Large, dark green leaves, divided into 13 to 25 oval leaflets per stalk

Sow fresh seed in pots in late summer or early autumn, using standard loam-based seed mix. Cover with mix, then a layer of coarse grit and place in a cold frame. Germination takes 4-12 weeks. Overwinter young seedlings in the cold frame. Plant out 2 years after germination.

Alnus *Betulaceae*

A genus of hardy deciduous trees and shrubs. Plant in any moist to waterlogged soil in a sunny situation.

Alnus incana (Alder, Gray alder) Zones 2–6
Small seeds
Deciduous hardy tree
Height up to 80 ft.
Male and female catkins are borne on the same tree in early spring
Oval, dark green leaves

Collect the seeds in late autumn. Put them in a plastic bag with a small amount of damp vermiculite. Label and place in refrigerator for a month.

Sow the chilled seed in winter in pots using standard loam-based seed mix. Cover with a layer of coarse grit and place under protection at 50°F. Germination takes 4-12 weeks. Overwinter young seedlings in the cold frame. Plant out 2 years after germination.

Arbutus *Ericaceae*

A genus of evergreen trees and shrubs, grown for their attractive bark and fruits. Plant in a fertile well-drained soil. Provide extra protection from strong, cold winds.

Arbutus andrachne (Strawberry tree) Zones 7–9
Medium seeds
Evergreen tree
Height up to 20 ft.
Clusters of urn-shaped cream/white flowers, followed by small, round, orange to red fruits
Flowers in late spring
Attractive peeling reddish bark. Oval, glossy, dark green leaves, which have a yellow green underside

Collect the fruit in autumn when they are red. Soak them for a minimum of 48 hours in hot, not boiling, water, this will soften the pulp from around the seed. Remove the pulp and sow seed immediately in pots using standard loam-based seed mix. Cover with a layer of coarse grit and place under protection at 60°F. Germination takes 4 to 12 weeks. If no germination occurs, place the container outside exposed to all weathers, or place in the refrigerator for 8 weeks, then return to a warm area. Overwinter young plants in a frost-free environment for 2 years then plant out.

Alternatively, store the cleaned seed in damp sand in the refrigerator and sow as above in the spring.

Betula *Betulaceae*

A genus of hardy deciduous trees and shrubs grown for their autumn colors and attractive bark.

Betula pendula (European white birch) Zones 2–6
Medium seeds
Hardy deciduous tree
Height up to 50 ft.
Purple/brown male catkins, pale green female catkins that stay on the tree until winter
Catkins appear in spring
Straight silver/white trunk, with pendulous branches
Triangular midgreen leaves with toothed edges

Collect the seeds in late autumn. Put them in a plastic bag with a small amount of damp vermiculite. Label and place in refrigerator for a month. Sow the chilled seed in winter in pots, using standard loam-based seed mix. Cover with a layer of coarse grit and place under protection at 50°F. Germination takes 4–12 weeks. Overwinter young seedlings in the cold frame. Plant out 2 years after germination.

Carica *Caricaceae*

A genus of tender evergreen trees and shrubs, minimum temperature 55°F. Plant in rich, loamy soil in sun or partial shade. In cool climates grow as a container plant.

Carica papaya (Papaya) Zone 10
Medium seeds
Evergreen tender tree
Height up to 20 ft.
Green-white flowers followed by large, pear-shaped fruit with yellow green skin, apricot pulp and round, black seeds in a central cavity
Flowers in summer
Large, palmate, lobed green leaves
Both male and female trees are needed for fruiting

Sow fresh seeds very thinly in early spring in pots, using standard loam-based seed mix, mixed with coarse horticultural sand. Mix to a ratio of 1 part soil mix + 1 part sand. Cover with perlite or vermiculite and place under protection at 75°F. Germination takes 3–4 weeks. Plant in growing position 2 years after germination.

If growing on as a container plant, use a standard loam-based potting mix, mixed with coarse horticultural sand. Mix to a ratio of 1 part mix + 1 part sand.

Carpinus

A genus of hardy deciduous trees grown for their autumn color. Plant in a fertile well-drained soil in sun or partial shade.

Carpinus betulus (Hornbeam) Zones 4–7
Medium seeds
Hardy deciduous tree
Height up to 80 ft.
Male flowers form in drooping catkins. Female flowers have crimson styles and green bracts grouped in shorter catkins. The fruit develops into triangular nutlets with three, long-lobed bracts.
The midgreen leaves are oval and pointed

Sow fresh seed in autumn in pots or cell packs, using standard loam-based seed mix. Cover with coarse grit, then place outside exposed to all weathers (see "Breaking Seed Dormancy", page 233, for more information). Germination occurs when the temperature reaches a constant 50°F. Grow on for 2 seasons before planting out.

Cedrus *Pinaceae*

A genus of hardy evergreen conifer. Plant in any soil, with the exception of waterlogged sites, in a sunny situation.

Cedrus deodara (Deodar Cedar) Zones 7–8
Medium seeds
Evergreen hardy tree
Height up to 80 ft.
The male flowers are erect, releasing yellow pollen in autumn. Female flowers are green. The cones are barrel-shaped and turn brown when ripe, which takes 2 years
Needle-like, gray green leaves

Collect the seeds in autumn from ripe, two-year-old cones. Break the wings off the seeds before putting them in a plastic bag with a small amount of damp vermiculite. Label and place in a refrigerator for 3 to 4 weeks. Sow the chilled seed in winter in pots, using standard loam-based seed mix. Cover with a layer of coarse grit and place under protection at 60°F. Germination takes 4–12 weeks. Overwinter young seedlings in the cold frame. Plant out 2 years after germination.

Cedrus deodara

Cercis *Caesalpiniaceae*

A genus of hardy deciduous shrubs and trees with attractive pea-like flowers. Plant in deep, well-drained soil in a sunny position.

Cercis siliquastrum (Judas tree) zones 6–8
Medium seeds
Hardy deciduous tree
Height up to 30 ft.
Clusters of bright pink flowers, which appear before the leaves, followed by long maroon/purple pods
Flowers in midspring
Heart-shaped, green leaves

Before sowing, rub the hard coated seeds gently with sandpaper, then soak for 24 hours in hot, not boiling, water. Sow seeds in autumn in pots or cell packs, using standard loam-based seed mix. Cover with vermiculite or soil mix and place under protection at 60°F.

Germination takes 3-4 weeks. Overwinter the young plants in a frost-free environment. Plant in growing position 2 years after germination.

Chamaecyparis *Cupressaceae*

A genus of hardy evergreen conifer. Plant in any soil, with the exception of waterlogged sites, in sun or partial shade.

Chamaecyparis lawsoniana (Lawson false cypress) zones 5–7
Medium seeds
Evergreen hardy tree
Height up to 80 ft.
Female flowers grow on the ends of small branchlets. Male flowers grow on the ends of branches and have black scales edged with white, becoming red when ripe. The cones are small and brown
Leaves are aromatic, dark green on top and lighter green below

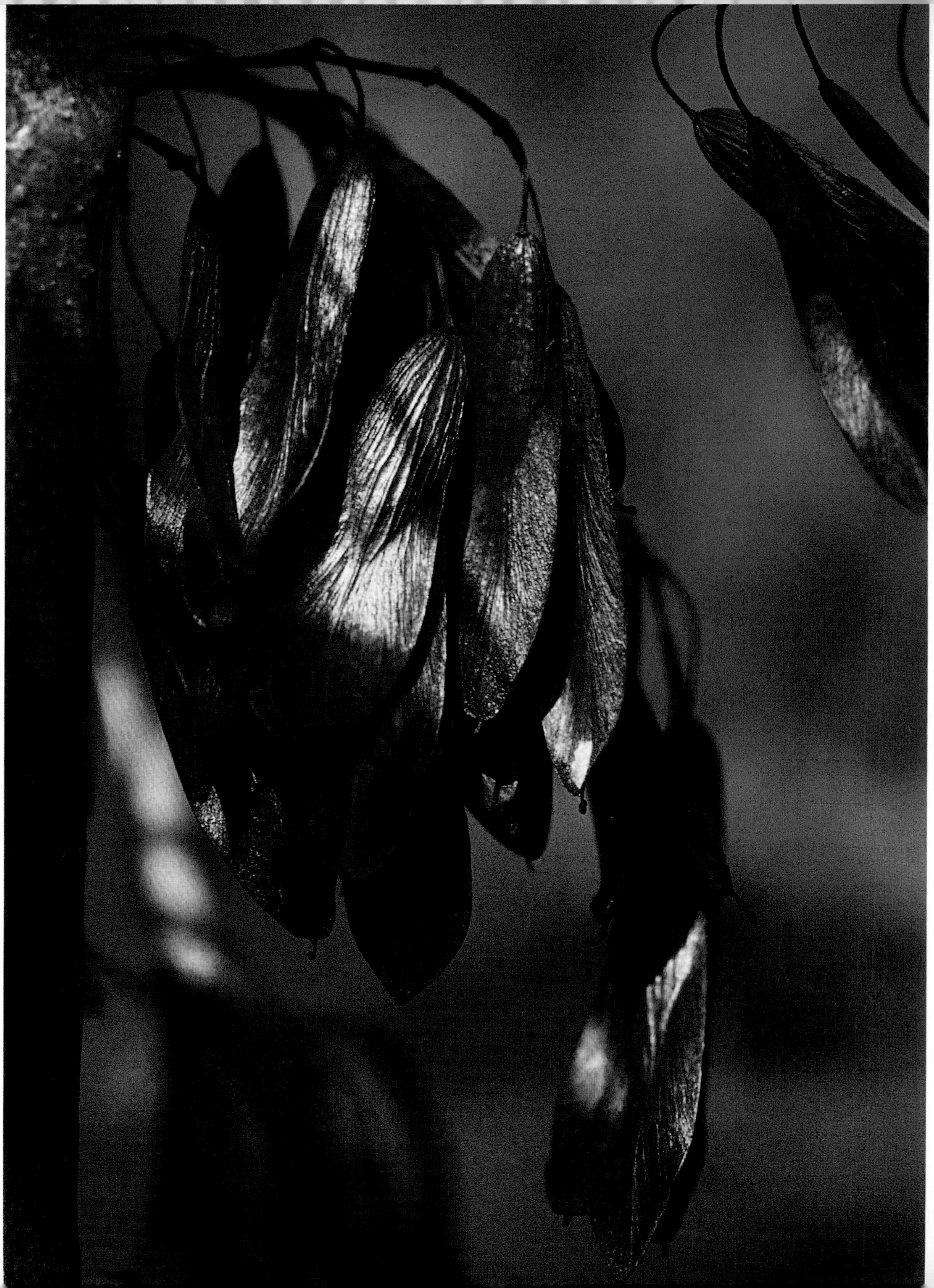

Fraxinus excelsior

Fraxinus *Oleaceae*

A genus of hardy deciduous trees and shrubs grown for their foliage. Plant in a fertile, well-drained soil, which does not dry out too much in summer, in a sunny situation.

Fraxinus excelsior (European ash) Zones 5–7
Medium seeds
Hardy deciduous tree
Height up to 100 ft
Male and female flowers often occur on the same tree on separate twigs, giving the tree a purple hue before the leaves come out in spring. Flowers are followed by bunches of single long-winged seed
Green leaves are divided into 9–13 leaflets

The seeds have a double dormancy. Collect them in autumn, and place in a plastic bag with a small amount of damp sand. Label and put in the refrigerator for 8 weeks. In winter, sow the chilled seed in pots, using standard loam-based seed mix. Cover with coarse grit. Place outside exposed to all weathers (see "Breaking seed Dormancy". page 233, for more information). Germination occurs after 2 winters outside. Plant in growing position 2 years after germination.

Genista *Papilionaceae*

A genus of half-hardy to fully deciduous shrubs and trees with pea-like flowers. Plant in fertile well-drained soil in a sunny situation.

Genista aetnensis AGM (Mount Etna broom) Zone 7
Medium seeds
Hardy deciduous tree
Height up to 25 ft.
Masses of fragrant golden yellow pea-like flowers
Flowers in midsummer
Bright green branches which are almost leafless

Before sowing in spring, scarify the fresh seeds using fine sandpaper (see page 233) then soak them in hot, not boiling, water. Discard any floating seeds and sow the remainder in pots, using standard soil-less seed mix, either peat or peat substitute mixed with 1/8in. fine grit. Mix to a ratio of 3 parts soil-less mix + 1 part fine grit. Cover lightly with perlite or vermiculite and place in a cold frame. Germination takes 2–5 weeks. Overwinter young plants in a cold frame. Plant in growing position 2 years after germination.

Ginkgo *Ginkoaceae*

A genus of hardy evergreen trees. Plant in a rich, fertile well-drained soil in a sunny position.

Ginkgo biloba AGM Zones 3–8
Large seeds
Hardy deciduous tree
Male and female flowers grow on separate trees. Males appear as small green catkins, females are small, stalked, round and knob-like, followed by oval fruit that look like small green plums
Fan-shaped, 2-lobed, midgreen leaves

Fertilisation of the seeds is by free-swimming male sperm which reach the ovules through a film of water. This is a method found in ferns, but in no other tree living today. In autumn, remove the pith from around the seed and then wash the seed in a mild detergent to remove any germination inhibitors.

Sow immediately in individual pots or cell packs, using standard loam-based seed mix mixed with coarse horticultural sand. Mix to a ratio of 1 part soil mix + I part sand. Cover with coarse grit and place in a cold frame. Germination takes 4–6 months, however it can take longer, so do not discard the container until the following spring. Plant in growing position 5 years after germination.

Ilex *Aquifoliaceae*

A genus of hardy to half-hardy evergreen and deciduous trees and shrubs grown for their berries and foliage. Plant in a well-drained soil in sun or partial shade.

Ilex aquifolium (English holly) Zones 6–8
Medium seeds (TOXIC SEED)
Hardy evergreen tree or shrub
Height up to 70 ft
Small, scented male and female flowers on separate trees. Round red berries on the female trees only
Flowers appear in late spring
Dark, glossy, spined, lobed leaves

In autumn, collect the fruit, clean the pith from the seeds (average 4 per fruit). Please note that although the seed is toxic, the fruit is not.

Sow immediately in pots or cell packs, using standard loam-based seed mix mixed with coarse horticultural sand. Mix to a ratio of 1 part soil mix + 1 part sand. Cover with coarse grit, water very well and place in a cold frame. Germination takes 4–6 months; however, it can take longer, so do not discard container until the following spring. Plant in growing position 2 years after germination.

Jacaranda *Bignoniaceae*

A genus of half-hardy deciduous or evergreen trees, minimum temperature 45°F. In warm climates, plant in fertile, well-drained soil in a sunny situation. In cool climates grow as a container plant.

Jacaranda mimosifolia Zones 9–10
Medium seeds
Half-hardy tree
Height up to 30 ft.
Trusses of vivid purple-blue flowers
Flowers in spring and early summer
Fast-growing, deciduous, rounded tree, with fern-like leaves

Sow seeds in early spring in pots or cell packs, using standard soil-less seed mix either peat or peat substitute. Cover with perlite or vermiculite. Place under protection at 68°F. Germination takes 4–6 weeks. It can be grown as a container plant for the conservatory, although it grows fast and will need clipping. Use a loam-based mix and ensure good drainage in the bottom of the pot.

vegetables, fruit & salads

There is nothing to beat homegrown vegetables. The information contained in this section has been tried and tested on my herb farm, or gleaned from fellow enthusiasts and from experts. Vegetable seed-counts in this chapter are approximate only, because exact counts depend on variety.

The most important element of a successful vegetable garden is the soil, and it must be fed correctly. In the majority of cases, vegetables need a soil that is free draining, rather than waterlogged, but does not dry out in summer. Where relevant, I have given an optimum growing temperature to act as a general guide. With this information to hand, your crops are sure to be successful.

Rheum rhaponticum, see page 224

Abelmoschus *Malvaceae*

Plant in a rich, open soil. It should be grown directly in the ground as the roots do not like being confined to a pot.

Abelmoschus esculentus (Hibiscus esculentus) "Clemson Spineless" (Okra, Gumbo)
Medium seeds: 50 per ounce
Half-hardy annual
Height up to 3–5 ft.
Yellow flowers with crimson centers
Heart-shaped, lobed, toothed leaves
Optimum growing temperature 60°–75°F

Soak seed for 24 hours prior to sowing in early spring, discarding any floating seeds. Sow in pots or cell packs, using standard soil-less seed mix, either peat or peat substitute. Cover with perlite or vermiculite, and place under protection at 68°F. Germination takes 5–10 days. When the seedlings are large enough to handle, plant out in a greenhouse bed 30 in. apart.
OR
Sow seeds under cover in late spring in prepared open ground, when the air temperature does not go below 15°C (60°F) at night. Germination takes 2–3 weeks. Thin seedlings to 20cm (8in) apart. Harvest at 8 weeks.

Allium *Alliaceae*

Plant in a rich soil that has been fed generously with well-rotted manure the previous autumn. It is worth letting a few plants of each species run to flower so that you can collect the seed in early autumn.

Allium cepa (Onion)
Medium seeds: 840 per ounce
Hardy annual
Height approx 24 in.
Optimum growing temperature 60°–75°F

Allium porrum (Leek)
Medium seeds: 784 per ounce
Hardy biennial
Height approx 2¹/₂ ft.
Optimum growing temperature 50°–60°F

Sow seeds from late winter to early spring in pots or cell packs, using standard soil-less seed mix, either peat or peat substitute. Cover with perlite or vermiculite, place under protection at 68°F. Germination takes 5–10 days.
OR
Sow seeds in late spring in prepared open ground, when the air temperature does not go below 55°F at night. Germination takes 2–3 weeks. Thin seedlings to 2–5 in. apart, depending on size of allium bulb.

Amaranthus *Amaranthaceae*

This is a vast group of plants, which are largely native to warm climates and need protecting from frost in cooler climates. The leaves of all the amaranths are edible and the seeds can be harvested for grain, but some are better suited to one of these purposes. All prefer to be planted in well-drained soil in sun or partial shade.

Amaranthus giganticus (Edible amaranth, Leafy amaranth)
Small seeds: 3,640 per ounce
Half-hardy annual
Height 1–2 ft.
Oval to heart-shaped, midgreen leaves

Sow seeds in early spring in pots or cell packs, using standard soil-less seed mix, either peat or peat substitute. Cover with perlite or vermiculite, place under protection at 68°F. Germination takes 5–10 days. Plant out 16 in. apart when all threat of frost has passed.

Apium *Apiaceae*

Plant in a deep, rich moist soil that has been mulched with well-rotted manure the previous autumn. Do not be tempted to use celery seed that has been bought for culinary use as it will not germinate.

Apium graveolens (Celery)
Small seeds: 2,400 per ounce
Hardy biennial
Height 2–3 ft.
Optimum growing temperature 60°–70°F

Sow seeds in early spring in pots or cell packs using standard soil-less seed mix, either peat or peat substitute. Cover with perlite or vermiculite, place under protection at 60°F. Germination takes 2–3 weeks. Plant out 15 in. apart after all threat of frost has passed.
OR
Sow seeds in late spring in prepared open ground, when the air temperature does not go below 55°F at night, as seedlings might bolt if temperature falls below 50°F. Germination takes 3–5 weeks. Thin out seedlings to 18 in. apart.
   Grow self-blanching varieties in blocks 12 in. square to encourage the stems to blanch naturally.

Apium graveolens var. *rapaceum* (Celeriac)
Small seeds: 7,000 per ounce
Hardy biennial
Height 12 in.
Optimum growing temperature 50°–70°F

Sow seeds in early spring in pots or cell packs, using standard soil-less seed mix, either peat or peat substitute. Cover with perlite or vermiculite, place under protection at 65°F. Germination takes 3–4 weeks. Plant out 12 in. apart in early summer, when all threat of frost has passed and the seedlings have been hardened off. Make sure that the little bulbous swellings at the bases of the plants stay at soil level. They must not be buried. Harvest in the autumn.

Brassica *Brassicaceae*

All brassicas like to feed well, although Brussels sprouts need more feeding than cabbages, and cauliflowers are positively greedy. However, do not become too enthusiastic and give them too much nitrogen, because this will promote soft growth and make the plants unable to survive the winter. Try to prepare your permanent beds the autumn before planting out, giving the site a good feed of well-rotted manure and compost. Traditionally, brassicas were all started off in outdoor seedbeds and then transplanted in permanent beds for growing on. However, if you have only a small garden and space is a luxury, you can sow in large cell packs outside and then transplant these into the garden.

Brassica oleracea Botrytis Group **(Cauliflower)**
Medium seeds: 910 per ounce
Hardy biennial
Height 24 in.
Optimum growing temperature 45°–70°F

Cauliflower can be sown throughout the growing season in cool climates. It is vital to choose the correct cultivar for the required cropping season. Consult current seed catalogs for the optimum sowing time for each variety.

In general, sow in late winter to early spring in pots or cell packs, using standard soil-less seed mix, either peat or peat substitute. Cover with perlite or vermiculite and place under protection at 68°F. Germination takes 6–10 days. Transplant in midspring or early summer to a permanent growing area. The distance between plants depends on the variety. Harvest approximately 8–10 weeks later.
OR
Sow in late spring to early summer in pots or cell packs—using standard soil-less seed mix, either peat or peat substitute—or in a seedbed. Cover with perlite or vermiculite. Place containers outside on a hard surface. Germination takes 6–10 days. Transplant in midsummer to a permanent growing area at a distance of 2 ft. apart. Harvest approximately 7–10 weeks later
OR
Sow seeds in late spring in prepared open ground, when the air temperature does not go below 50°F at night. Germination takes 2–3 weeks. Thin seedlings to 2 ft. apart.

Brassica oleracea var. *capitata* **(Cabbage)**
Medium seeds: 910 per ounce
Hardy biennial grown as an annual
Height spring cabbage 10–12 in., winter cabbage 16 in.
Optimum growing temperature 15–18°C (60–65°F)

Cabbage can be sown throughout the growing season in cool climates; it is vital to choose the correct cultivar for the required cropping season. Consult current seed catalogs for the optimum sowing time for each variety.

Sow in late winter to early spring in pots or cell packs, using standard soil-less seed mix, either peat or peat substitute. Cover with perlite or vermiculite and place under protection at 68°F. Germination takes 6–10 days. Plant out, after a period of hardening off, 15 in. apart.

OR
Sow in late spring to early summer in pots or cell packs, using standard soil-less seed mix, either peat or peat substitute. Cover with perlite or vermiculite and place outside on a hard surface. Germination takes 6–10 days. Plant out at a distance of 15 in. apart.
OR
Sow seeds in late spring to early summer in prepared, open ground, when the air temperature does not go below 50°F at night. Germination takes 2–3 weeks. Thin seedlings to 15 in. apart. Do not allow seedlings to dry out during hot spells.

Brassica oleracea Gemmifera Group **(Brussels sprouts)**
Medium seeds: 910 per ounce
Hardy biennial grown as an annual
Height 3¹/₂ ft.
Optimum growing temperature 45°–70°F

Brussels sprouts must have a fertile soil; however, it is better if the soil has been well manured for a previous crop because Brussels sprouts need a firm soil in which to anchor themselves. Dig the site as early as possible, to allow the soil to settle.

Sow in early to late spring in pots or cell packs, using standard soil-less seed mix, either peat or peat substitute. Cover with perlite or vermiculite and place under protection at 68°F. Germination takes 6–10 days. Plant out after a period of hardening off 18–24 in. apart, depending on variety.
OR
Sow seeds in late spring to early summer in prepared open ground, when the air temperature does not go below 50°F at night. Germination takes 2–3 weeks. Thin seedlings to a distance of 18–24 in. apart, depending on variety.

Recipe for sowing, planting, and growing **Brassica oleracea Gemmifera Group**
Brussels sprouts are one of the great vegetables of autumn and winter, if picked at the right moment when they are small and firm. For the best flavor, do not overcook.
Ingredients
3 seeds per cell or 7 seeds per pot
1 flat with cell packs
OR
1 x 4 in. pot
Standard soil-less seed mix, either peat or peat substitute
Fine-grade perlite (wetted) or vermiculite
White plastic plant label
OR
A prepared site in the garden

Method Fill the cell packs or pots with soil-less mix, smooth over, tap down and water in well. Place 3 seeds per cell or 5 seeds per pot, spaced equally on the surface of the mix. Press gently into the mix with the palm of the hand. Cover the seeds with fine-grade perlite (wetted) or vermiculite and label with the plant name and date. Place the pot in a warm light place, out of direct sunlight, at an optimum temperature of 68°F. Restrict watering to a minimum until germination has taken place, after 6–10 days. The seedlings will be ready to plant out or pot

213

Capsicum annum Longum Group

up approximately 2–3 weeks after germination. Plant out, after a period of hardening off, 18–24 in. apart, depending on variety, into their permanent growing position. Firm the young plants in well and protect from birds as they love sprouts.

OR

Sow directly in a prepared site in the garden when the nighttime temperature does not fall below 50°F. Sow the seeds thinly

OR

For later crops, sow seeds in late spring or early summer in prepared, open ground in rows 18 in. apart, when the air temperature does not go below 50°F at night. Germination takes 2–3 weeks. Thin seedlings to 12–18 in. apart, depending on variety.

Brassica oleracea Italica Group **(Sprouting broccoli)**

Ipomoea batatas **(Sweet potato)**
Medium seeds
Half-hardy perennial grown as an annual
Height 18 in
Pretty white flowers with dark purple/maroon centers
Midgreen large leaves

Sow seeds in spring in pots or cell packs, using standard soil-less seed mix, either peat or peat substitute. Cover with perlite or vermiculite and place under protection at 75°F. Germination takes 10–20 days. Plant out after a period of hardening off and all threat of frost has passed at a distance of 12 in apart.

Lactuca *Asteraceae*

Lettuce is a cool-climate plant, growing best at temperatures below 70°F. It is a fast-growing vegetable, so plant in a very fertile soil that has been well fed the previous autumn and which retains moisture throughout the growing season. If grown in too high a temperature, Lettuce will bolt and taste bitter and, in cold damp weather, it is prone to disease and slugs.

Lactuca sativa **(Lettuce)**
Medium seeds: 1,456 per ounce
Height 3–12 in. depending on variety
Optimum growing temperature 50°–68°F

There are many types of lettuce, varying in color from green to rust, with leaves of all shapes and sizes, from round to serrated. Lettuce can be sown throughout the growing season in cool climates. It is vital to choose the correct cultivar for the cropping season, so consult current seed catalogs for the optimum sowing time for each variety.

Sow seeds in late spring in prepared open ground in drills, when the night-time temperature does not go below 50°F. Germination takes 10–14 days. Thin seedlings to 6–12 in depending on variety.

OR

Sow seeds in spring and autumn in pots or cell packs, using standard soil-less seed mix, either peat or peat substitute. Cover with perlite or vermiculite and place in a cold frame. Do not allow the young plants to get too hot, ventilation is important. Germination takes 8–14 days. Plant out, after a period of hardening off, at a distance of 6–12 in., depending on variety.

Lepidium *Brassicaceae*

Cress prefers growing in cool, moist conditions, so plant in a fertile, moist soil. If grown in conditions that are too hot, it will bolt and go to seed very quickly.

Lepidium sativum **(Cress, Garden cress, Pepper cress)**
Medium seeds: 1,120 per ounce
Half-hardy annual
Height 3–6 in.
Deeply curled, deep green leaves with a spicy flavor.

Sow seeds in late spring in prepared open ground, when the air temperature does not go below 50°F. at night. Germination takes 2–4 days. Thin seedlings to 3/4 in. apart so that they are not overcrowded.

Lycopersicon *Solanaceae*

Tomatoes are native to the tropics, where they are perennials. In gardens, they are grown as annuals. Plant in a rich, moist soil in a sunny position.

Lycopersicon esculentum **(Tomato)**
Half-hardy perennial grown as an annual
Medium seeds: 980 per ounce
Height 3–4 ft. for bush varieties, up to 7 ft. for indeterminate varieties
Optimum growing temperature 70°–75°F
Small yellow flowers in summer followed by red fruit
Midgreen leaves. Some cultivars have serrated leaves, some lobed

Sow seeds in spring in pots or cell packs, using standard soil-less seed mix, either peat or peat substitute. Cover with perlite or vermiculite and place under protection at 70°F. Germination takes 5–10 days. Plant out, after a period of hardening off and when all threat of frost has passed, at a distance of 18–30 in. depending on variety.

OR

Sow seeds in late spring in prepared open ground, when the air temperature does not go below 45°F at night. Germination takes 10–14 days. Thin seedlings to a distance of 18–30 in. depending on variety.

Pastinaca *Apiaceae*

Parsnip is a cool-season crop. Plant in a deep, fertile, light, free-draining soil in sun or partial shade.

Pastinaca sativa **(Parsnip)**
Medium seeds: 762 per ounce
Hardy biennial grown as an annual
Height 15 in.
Small, yellow-green flowers in second season, followed by golden-yellow seed heads
Midgreen, lobed leaves

For good germination it is most important that the seed is fresh.

Sow the seeds thinly in spring for autumn and early winter crops in prepared, open ground, 3/4 in. deep, when the soil temperature is above 50°F. Germination should take 10–14 days. In lower temperatures, it is very slow. Be careful when thinning the plants out to 4 in. apart not to attract the carrot-root fly. The best prevention is to water the site well, thin the plants in the early evening and cover with floating row covers for a few days.

OR

Sow the seeds in early autumn for early spring crops. Sow in prepared open ground 3/4 in. deep and 4 in. apart. Germination takes 10–14 days. This is a good option if you live in a warm climate, but is not recommended in very cold climates.

Phaseolus *Papilionaceae*

Prepare the ground thoroughly for beans, with plenty of well-rotted compost. The soil should be fertile, moisture retentive and free draining, and the best situation is sunny and sheltered from strong winds to encourage pollination by bees.

Phaseolus coccineus (Runner beans)
Very large seeds: 3 per ounce (approximately)
Half-hardy perennial grown in cool climates as an annual
Height 10 ft. or more

Seeds do not germinate below 54°F so, if you want an early crop, start sowing seeds singly in early spring in pots or cell packs using standard soil-less seed mix, either peat or peat substitute. Cover with perlite or vermiculite and place under protection at 65°F. Germination takes 7–10 days. Plant out, after a period of hardening off, at a distance of 6 in.
OR
Sow seeds in late spring in prepared, open ground at a depth of 2 in., 6 in. apart, when the air temperature does not go below 45°F at night. Germination takes 2–3 weeks.

Phaseolus vulgaris (Bush beans and Pole beans)
Large seeds: 8–30 per ounce
Half-hardy annual
Height up to 18 in. for bush forms, climbers up to 10 ft.
Green beans prefer a lighter soil than runner beans

Sow seeds singly, in early spring, in pots or cell packs, using standard soil-less seed mix, either peat or peat substitute. Cover with perlite or vermiculite and place under protection at 65°F. Germination takes 5–10 days. After hardening off, plant out at a distance of 6–9 in., depending on variety. Harvest from midsummer onwards.
OR
Sow seeds in late spring, at a depth of 2 in. and 6–9 in. apart, depending on variety. Sow in prepared, open ground, when the air temperature does not go below 45°F at night. Germination takes 2–3 weeks. Harvest from early autumn onwards.

Phaseolus coccineus

Pisum *Papilionaceae*

Peas are quite fussy; they like a moisture-retentive, free-draining soil, which should be fed with well-rotted compost in autumn. They suffer in both cold, wet soil and drought conditions.

Pisum sativum (Garden peas)
Large seeds: 8–17 per ounce
Annual
Height 18 in.–6 ft. depending on variety
Optimum growing temperature 55°–64°F
Small white, pink, or purple flowers
Small oval leaves

The seeds will not germinate in a soil below 45°F or in hot conditions.

Sow in spring as soon as the soil temperature reaches 50°F. Before sowing, soak the seeds overnight in warm water. Sow at a depth of 2 in. and 2 in. apart in a prepared site, or in a length of plastic rain gutter (see recipe opposite). Germination takes 7–14 days.
OR
If you live in a cold area, and want to start your crop off as soon as possible, sow seeds singly in early spring in pots or cell packs, using standard soil-less seed mix, either peat or peat substitute.
   Place under protection at 65°F. Germination takes 5–10 days. Plant out when all threat of frost has passed at a distance of 2 in.

Pisum sativum var. *saccharatum* (Snap peas, Sugar peas)
Large seeds: 8–17 per ounce
Annual
Optimum growing temperature 55°–65°F
Small, white or red flowers
Oval, pointed, green leaves

The seeds will not germinate in a soil below 45°F, nor in hot conditions. Therefore, sow in spring as soon as the soil temperature reaches 50°F. Before sowing, soak the seeds overnight in warm water. Sow at a depth of 2 in. and 2–3 in. apart in prepared, open ground. Germination takes 7–14 days.
OR
If you live in a cold area and want to start your crop off as soon as possible, sow the seeds in early spring in pots or cell packs, using standard soil-less seed mix, either peat or peat substitute. Cover with perlite or vermiculite and place under protection at 65°F.
   Germination takes 5–10 days. Plant out when all threat of frost has passed at a distance of 2–3 in.

Raphanus *Brassicaceae*

Grown slowly, radishes have a strong flavour and can be chewy. Grown in a rich, moisture-retentive soil they will be sweet and crisp.

Raphanus sativus (Radish)
Medium seed: 224–504 per ounce
Hardy and half-hardy biennial grown as an annual
Height 4–18 in, depending on variety
Small, pink white flowers in summer, depending on variety
midgreen leaves

Recipe for sowing, planting, and growing **Pisum sativum**
Peas have been found on an archaeological dig dating back to 9,750 B.C. They did not become fashionable until the seventeenth century. They are an immensely versatile vegetable; both the pod and the peas are edible and great as a vegetable or as a soup.

Ingredients
1 seed per pot or cell
Small bowl
Warm water
1 flat with cell packs
OR
1 x 4 in. pot
Standard soil-less seed mix, either peat or peat substitute
White plastic plant label
OR
Length of plastic rain gutter 3–6 ft long
Standard soil-less seed mix, either peat or peat substitute
2 white plastic plant labels

Method In spring fill a bowl with warm water, add the seeds to the water and soak for 12 hours. Fill the cell packs or pots with soil-less mix, smooth over, tap down and water in well. Put 1 seed in each cell or pot. Press the seed gently into the mix until it is totally buried, water lightly again, then label with the plant name and date. Place the pot or cell packs in a warm, light place, out of direct sunlight ,at an optimum temperature of 65°F. Keep watering to a minimum until germination has taken place, which takes 5–10 days. Once germinated, move the seedlings to a cooler environment, then slowly harden off the young plants, before planting out in a prepared site when all threat of frost has passed. Plant out at a distance of 2 in.
OR
When the nighttime temperature is about 50°F, fill a length of plastic rain gutter with soil-less mix, 1/2 in. from the rim. Sow the pre-soaked pea seeds in double rows about 2i n. apart. Water them to settle the mix and cover the seeds with more mix, filling the gutter up to the rim. Give a further, light watering and label. Place the gutter on a sunny windowsill or in a sheltered place to germinate. When the seedlings are 3–4 in tall they can be planted in the garden. Dig a shallow trench the same length and depth as the gutter. Ease the well-rooted seedlings, just as they are, into the trench gently. Firm in well, water and label. If you are having difficulty, you can divide the pea seedlings into 12 in. lengths then plant them out in the trench a section at a time.

Radishes can be sown throughout the spring and early summer; it is vital to choose the correct cultivar for required sowing time. So consult current seed catalogs for the optimum sowing time for each variety. There are many to choose from—small round, small long, large and all the oriental ones.

Sow seeds thinly in late spring in prepared open ground, when the air temperature does not go below 45°F at night. Germination takes 5–10 days. Thin seedlings to 2–6 in. apart, depending on variety.

Rheum *Polygonacaea*

Rhubarb should be grown in a moisture-retentive soil that has been fed with well-rotted manure or compost in the autumn in sun or light shade. Once established it needs winter cold to bring it out of dormancy. For this reason, it does not do well in warm climates.

Rheum rhaponticum (Rhubarb)
Medium seeds: 280 pr ounce
Hardy perennial
Height 3–5 ft.
Frothy, small, cream flowers in early summer which should be cut off to stop the plant becoming weak
Large green leaves.

Sow seeds in early spring in pots or cell packs using standard soil-less seed mix, either peat or peat substitute. Cover with perlite or vermiculite and place under protection at 65°F. Germination takes 4–14 days. Plant out, when all threat of frost has passed, at a distance of 3 ft. apart.
OR
Sow seeds thinly in late spring in prepared, open ground, when the air temperature does not go below 50°F. at night. Germination takes 10–20 days. Thin seedlings in two sessions, first to 12 in. apart, then to 3 ft. apart.
   Seed-grown rhubarb can be harvested gently from the second year's growth.

Rorippa *Brassicaceae*

Plant in a very moist soil at the waters edge, in a shady position. If you do not have a pond, line a container with plastic before filling with potting mix, this will help to keep the mix moist. Place the container in a shady position.

Rorippa nasturtium aquaticum (*Nasturtium officinale*) (Watercress)
Small seeds: 16,800 per ounce
Hardy perennial
Height 4–24 in.
Clusters of tiny white flowers in summer, followed by seed pods.
Dark green, lobed, strong-tasting leaves.

Sow seeds in early spring in pots using standard soil-less seed mix, either peat or peat substitute. Cover with perlite or vermiculite and place under protection at 65°F. Germination takes 5–10 days. Keep well watered during germination. Once germinated, continue to water frequently. Plant out as soon as all threat of frost has passed. Plant either in a calm part of an unpolluted stream or in a container, as described in the introduction to this entry, at a distance of 6 in apart.

Rumex *Polygonaceae*

Sorrel will tolerate any soil, but thrives in a damp, fertile soil in sun or partial shade. Both varieties mentioned below are known as French sorrel which is confusing.

Rumex acetosa (Broad-leaved sorrel, French sorrel)
Small seeds: 2,800 per ounce
Hardy perennial
Height up to 4 ft when in flower
Boring rust-colored flowers in summer
Long, oval, midgreen leaves

Rumex scutatus (Buckler leaf sorrel, French sorrel)
Small seeds: 2,800 per ounce
Hardy perennial
Height 20 in.
Small, green-red flowers in summer
Shield-shaped, midgreen leaves

Sow seeds in early spring in pots or cell packs, using standard soil-less seed mix, either peat or peat substitute. Cover with perlite or vermiculite and place under protection at 60°F. Germination takes 5–10 days. Plant out, after a period of hardening off, at a distance of 12 in.
OR
Sow seeds thinly in late spring in prepared open ground, when the air temperature does not go below 45°F at night. Germination takes 2–3 weeks. Thin seedlings to 12 in. apart.

Sium *Apiaceae*

Skirret prefers a sunny site and a soil that has not been manured recently. It is quite happy to follow a crop that was fed and survive on what is left, as long as the soil is light and free draining.

Sium sisarum (Skirret)
Medium seeds: 2,240–2,800 per ounce
Hardy perennial usually grown as an annual
Height up to 3 ft when in flower
Umbels of white flowers in late summer on second year's growth
Sharply toothed oval leaves

Sow seeds thinly in prepared open ground, from early spring for autumn harvest or autumn for late spring harvest, in drills 1 in. deep. Germination takes 2–3 weeks, or longer, depending on the time of year. Thin seedlings to 10 in. apart. Roots take about 4 months to mature.

Solanum *Solanaceae*

Eggplants and potatoes should be planted in a fertile, deep, free-draining soil in a sunny position.

Solanum melongena (Eggplant)
Medium seeds: 560 per ounce
Half-hardy perennial grown as an annual
Height 12–30 in.
Optimum growing temperature 77°–86°F plus high humidity

Small, mauve flowers, followed by fruit in various shapes and colors
Large, oval, soft green leaves

Before sowing the seeds soak them for 12 hours, then sow in early spring in pots or cell packs using standard soil-less seed mix, either peat or peat substitute. Cover with perlite or vermiculite and place under protection at 70°F. Germination takes 7–14 days. Plant out after a period of hardening off, when all threat of frost has passed, at a distance of 24in.

OR

Grow on as a container plant, use a standard, loam-based potting mix in a 8–12 in. pot.

Recipe for sowing, planting, and growing Solanum melongena
Ingredients
3 seeds per cell or 5 seeds per pot
Small bowl
Warm water
1 flat with cell packs
OR
1 x 4 in pot
Standard soil-less seed mix, either peat or peat substitute
Fine-grade perlite (wetted) or vermiculite
White plastic plant label

Method Before sowing the seeds soak them for 12 hours, then fill the cell packs or pots with soil-less mix, smooth over, tap down and water in well. Sow 3 seeds per cell or 5 seeds per pot. Cover with fine-grade perlite (wetted) or vermiculite. Label with the plant name and date. Place the pot in a warm light place, out of direct sunlight, at an optimum temperature of 70°F. Germination takes 7–14 days. When the seedlings show, move them to a cooler environment, 65°F by day and no less than 60°F by night. Make sure they have plenty of light but avoid direct sunlight. Keep the seedlings moist. They will be ready to plant out or pot up when they are 3–4 in. in height. Plant out at 24 in. or grow on as a container plant using a standard loam-based potting mix in a 8–12 in. pot. They combine well with cucumbers.

Solanum tuberosum (Potato)
Small seeds: 4,900 per ounce
Half-hardy perennial grown as an annual
Height up to 30 in. and taller for some varieties
Optimum growing temperature 60°–64°F
Flowers can vary in color from mauve, purple, pink to cream.
Midgreen leaves

Potatoes can be grown from seed. The tubers that are produced are smaller and their skin is thin, so they make a very good summer crop. However, in cool climates with a short growing season, seed potatoes are often chitted under cover, to give the plant a good start.

Sow seeds in early spring in pots or cell packs using standard soil-less seed mix, either peat or peat substitute. Cover with perlite or vermiculite and place under protection at 60°F. Germination takes 7–14 days. Plant out when all threat of frost has passed, at a distance of 12 in.

Spinacia *Chenopodiaceae*
Plant in a fertile moisture-retentive soil in partial shade. Spinach is notoriously prone to bolt if grown in hot, dry conditions.

Spinacia oleracea (Spinach)
Medium seeds: 252 per ounce
Hardy annual
Height up to 12 in.
Long, oval, pointed leaves

Sow seeds in early spring in pots or cell packs using standard soil-less seed mix, either peat or peat substitute. Cover with perlite or vermiculite and place under protection at 64°F. Germination takes 5–10 days. Plant out, after a period of hardening off, at a distance of 6 in. apart.
OR
Sow seeds thinly in late spring or early autumn for hardy varieties in prepared, open ground at a depth of 3/4 in., when the air temperature does not go below 45°F at night. Germination takes 2–3 weeks. When the seedlings are large enough, thin to 6 in apart.

Tragopogon *Asteraceae*
Plant in a light, free-draining, fertile soil and a sunny position.

Tragopogon porrifolius (Salsify)
Medium seeds: 840 per ounce
Hardy biennial grown as an annual
Height up to 4 ft. in flower
Pretty mauve flowers in summer of second year (the flower buds are edible)
Grass-like leaves

Always use fresh seed. Sow seeds in early spring in pots or cell packs using standard soil-less seed mix, either peat or peat substitute. Cover with perlite or vermiculite and place under protection at 60°F. Germination takes 7–14 days. Plant out when all threat of frost has passed at a distance of 4 in.
OR
Sow seeds thinly in late spring in prepared, open ground at a depth of 3/4 in, when the air temperature does not go below 45°F at night. Germination takes 2–3 weeks. When seedlings are large enough, thin to 4 in apart.

Valerianella *Valerianaceae*

Plant in any site, with the exception of waterlogged soil, in sun or light shade. Will not grow well in hot climates.

Valerianella locusta (Corn salad, Lamb's lettuce)
Small seeds: 3,220 per ounce
Hardy annual
Height 4 in
Spoon-shaped leaves, color and size dependent on cultivar, midgreen to dark green

Sow seeds in late spring in pots or cell packs, using standard soil-less seed mix, either peat or peat substitute. Cover with perlite or vermiculite and place under protection at 58°F. Germination takes 4–5 days. Plant out after a period of hardening off, when there is no threat of frost, at 4 in apart.
OR
Sow seeds thinly in midsummer in prepared, open ground. Germination takes 7–10 days. Thin seedlings to 4 in. apart. Do not allow late sowings to dry out. Water well.

Vicia *Papilionaceae*

Broad beans prefer a free-draining, well-manured soil in a sunny position.

Vicia faba (Broad beans or fava beans)
Very large seeds: 3 per ounce (approx.)
Half-hardy and very hardy annual
Height 1–4 ft. depending on variety
White flowers in summer
Gray green oval leaflets

Sow seeds in autumn or early spring in pots or cell packs, using standard soil-less seed mix, either peat or peat substitute. Cover with perlite or vermiculite and place under protection at 50°F, or outside if the night temperature does not fall below 45°F. Germination takes 1–2 weeks. Protect seedlings from frost. Plant out after a period of hardening off at a distance of 10–12 in depending on variety.
OR
Sow seeds in late spring in prepared open ground 5cm (2in) deep, 10cm (4in) apart, when the air temperature does not go below 5°C (40°F) at night. Germination takes 2–3 weeks.
The main harvest time for broad beans is summer. If you have old plants left, dig them into the garden as they make a very good green manure, fixing nitrogen in the soil. They also give bulk to compost.

Xanthophthalmum *Asteraceae*

Plant in a fertile, well-draining soil in sun or partial shade.

Xanthophthalmum coronarium (Chopsuey green, Chrysanthemum greens)
Medium seeds: 1,176 per ounce
Hardy annual
Height 6 in. in the green, 3 ft. in flower
Yellow, yellow and white, daisy-like flowers in summer
Serrated, spoon-shaped leaves.

If you want to grow this plant as a crop, pick regularly to delay the plant running to seed and becoming tough.

Sow seeds in early spring in pots or cell packs, using standard soil-less seed mix, either peat or peat substitute. Cover with perlite or vermiculite and place under protection at 60°F. Germination takes 7–10 days.
OR
Sow seeds in late spring in prepared open ground, when the soil temperature is 45°F at a distance of 8 in in drills 12 in apart. Germination takes 2–3 weeks

Zea *Poaceae*

Plant in full sun, in a free-draining site fed with well-rotted manure. Once the ears start to form, water regularly.

Zea mays var. *saccharata* (Sweet corn)
Large seeds: 14 per ounce
Half-hardy annual
Height 4–9 ft. depending on variety
The male flowers grow in spikes, the soft silks come in shades of silver, gold, bronze, and red
Long, lance-shaped, green leaves

Sweet corn hates being transplanted, so it is best to grow it in its final growing position. Sow the seeds in pairs (thin the weaker seedlings later) in late spring, in prepared, open ground, when the soil temperature is 50°F) Sow at a distance of 16 in. apart in drills 24 in. apart. Germination takes 1–2 weeks but can be faster in warmer weather.
OR
If you live in a cold climate, sow single seeds in early spring in cell packs, using standard soil-less seed mix, either peat or peat substitute. Cover with perlite or vermiculite and place under protection at 60°F. Germination takes 7–10 days.
Transplant to its growing position, as soon as there is no threat of frost, at a distance of 16 in. apart.

practical information

how plants reproduce

Before growing from seed, it is helpful to understand how plants work. Understanding the process, as it occurs naturally, helps the gardener to reproduce the correct growing conditions in the garden.

Plant structure

In its simplest form, every plant is made up of four essential parts:

The roots anchor the plant firmly in the growing medium (usually soil) and prevent it from being blown over by the wind. Roots absorb water, nutrients and mineral salts from the soil and pass them into the stem. Roots frequently act as a food store.

The stem is the superhighway of the plant. It supports the shoots, spaces out the leaves so that they receive adequate air and sunlight, conducts water from the soil to the leaves, and food from the leaves to other parts of the plant. In some cases, it also holds the flowers above ground, thus assisting in pollination. Photosynthesis might also occur in some green stems.

The leaves are, in most plants, the primary organs of photosynthesis, the process by which the chlorophyll in a plant captures light energy and converts it into chemical energy. The plant needs a good supply of water and carbon dioxide from the air in order to achieve this.

The flower produces seed, ensuring that there will be new plants in the future. Some plants have bisexual flowers. These contain both stigmas and stamens. Others have separate male and female flowers. Dioecious plants have flowers of only one sex borne on each plant.

The female part of the flower, called the pistil, is made up of the stigma, the style and the ovary. The stigma is at the tip of the style, a tube that connects it to the ovary. Pollen is deposited on the surface of the stigma during pollination. The ovary, which is usually hollow, varies in shape, size and color. Ovules are attached to the walls of the ovary. It is these ovules that develop into the seeds once they have been fertilized by pollen from the male parts of a flower.

The ovary, style, and stigma form the carpel. There could be several carpels per flower, or only one. The carpel is always positioned in the center or apex of the flower, giving the maximum chance of pollination.

The male part of the flower consists of the stamens. Each stamen is composed of a stalk or filament and an anther, which contains and dispenses the pollen. It is the pollen that fertilizes the female parts and turns the ovules into viable seed.

Pollination

Before a plant can produce seed, the flower of the plant must first be pollinated.

Animals, birds, bees, and the wind are all used as vehicles for transporting the pollen from one plant on to the stigma of another. This begins the process of fertilization. Cross-fertilization often produces healthier and more viable seed than self-fertilization. The majority of plants, especially wild species, have systems to prevent self-pollination. However, in nature there are always exceptions to the rule and self-fertilization does sometimes occur.

Fertilisation

Once the pollen is deposited on the stigma, it absorbs the sugar from the syrupy liquid on the surface of the stigma.

The pollen swells and eventually grows a pollen tube, which is hollow but contains the three essential nuclei necessary for fertilzation. The tube grows into the style and all the way along it to reach the ovary. Once it reaches the ovary, one of the nuclei, the vegetative one, disappears, and the remaining two travel on and reach the ovum. Here one of the nuclei fuses with the ovum nucleus. At this stage fertilization is accomplished and the ovum becomes a live embryo inside the ovule, which now develops into a viable seed. The petals of the flower begins to fade and fall and the ovary begins to swell.

Seed dispersal

Once seeds have matured they must be dispersed.

If all the seeds fell in one place, they would have to compete for water, light and nutrients. Many seeds are surrounded by flesh, which could be soft, or hard and dry. These seed cases might provide food for insects, animals, or humans. In this way, the seeds are carried great distances before being excreted. The digestive processes of certain animals boost germination. Some seeds actually rely on this process to break their dormancy. Other plants have seeds encased in burrs, or attached to thorns or hooks, which are carried by animals who brush by the parent plant. Other seeds are winged and are carried by the wind. These plants produce very light seeds with plumes or feather-like parachutes, making them dispersed easily over long distances by the wind. Some plants which grow by water produce seeds or fruits that are waterproof and buoyant.

Hybridization

At its simplest, a hybrid is a cross between two different plants.

Commercial growers spend a lot of time and money creating new, stable hybrids that can be relied upon not to revert to the form and color of one or other of the parent plants. With patience, the amateur gardener can produce some very attractive hybrids. There is a number of plants that hybridize naturally in the garden, such as *Aquilegia* and *Iris*. Natural hybrids are rarely stable, so plants grown from their seed are usually different from the parent plant.

To create a hybrid, select two stable parent plants. Choose one to be the female (seed producer) and one to be the male (pollen producer). Remove the stamens from the female plant to make quite sure it will not self-pollinate. With the pollen from the male plant, hand pollinate the stigma on the female plant, then cover the female flower with a muslin bag to protect it from contamination by bugs or wind until the seeds are developed.

The first generation of plants is called F1 hybrids. If they, in turn, are crossed with themselves, the second generation is called F2 hybrids. It is worth remembering that seeds from F1 hybrids are rarely worth saving. If sown, they will almost invariably produce plants inferior to the parent plants.

Buddleja davidii, see page 177

Soaking

Some seeds benefit from a period of soaking in hot water.

Baptisia and *Laburnum* are both examples of this seed type. Place seeds in a bowl and pour over them 4–5 times their volume of very hot, not boiling, water. Soak for 24 hours or until they swell, discarding any floating seeds as unviable. As soon as the seed swells. it must be sown before it dries out. If seeds do not swell, scarify them either by pricking the seed all over (avoiding the hilum) or by rubbing them with sandpaper. A good way to do this is to line a jam jar with fine sandpaper, put the drained seeds into the jar and shake from side to side so that they are abraded gently by the sandpaper, then soak them again. If soaking for longer than 24 hours, change the water every day.

Seeds that contain natural inhibitors to induce dormancy also benefit from soaking, as this will leach out the inhibitor. In some cases, the seed will also need washing with a mild detergent to remove the inhibitor. This technique is helpful when germinating *Gingko* seed.

Another method of soaking seed, particularly suitable for sweet peas, is to soak two or three layers of paper towels on a plate, space the seeds evenly on the towels and cover with two or three further layers of pre-soaked paper towels. Keep moist and inspect daily by lifting the corner of the paper. As soon as the seeds begin to sprout they can be transplanted.

Fire and smoke

In nature, some seeds germinate only after a bush fire.

Many of the seeds requiring this treatment originate in South Africa and Australia. Trying to replicate these conditions in the domestic environment can get a bit too exciting! However, you can buy kits containing smoke water and sheets of paper that have been smoke treated. Seeds are sown directly on to the smoke-treated paper and allowed to germinate in the usual way. If you are feeling brave, you can make your own tent-like contraption over the flats or pots, light a fire of damp bracken and green sticks, guide the smoke into the tent and leave the seeds in the smoke for 3–24 hours. Alternatively,

you can cover a flat containing the seeds with dry bracken, set it alight and then water in the ashes.

For more detailed instructions, please refer to specific plants in the book. Here you will find instructions regarding extra heat, the need to mix different temperatures and the types of soil mix to use.

Seed

Whether you choose to use home-harvested seed or bought seed, you will derive great pleasure from growing your own plants. Organic gardeners should avoid chemically treated seed.

Untreated or natural seeds Used by organic gardeners, these seeds have been harvested, dried and cleaned. They will have received no other treatment.

Primed seeds These seeds have been treated, so that they germinate quickly. Follow the instructions on the seed packet.

Chitted seeds Chitted or pregerminated seeds aid the gardener to an early start. They are usually sold in plastic containers and should be sown immediately, following the instructions on the container.

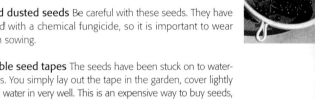

Pelleted seeds These seeds are encased in a coating or pellet that makes them easier to handle. They are also easier to sow evenly. It is crucial to water in the seeds very well after sowing, in order to dissolve the pellets.

Coated and dusted seeds Be careful with these seeds. They have been treated with a chemical fungicide, so it is important to wear gloves when sowing.

Water-soluble seed tapes The seeds have been stuck on to water-soluble tapes. You simply lay out the tape in the garden, cover lightly with soil and water in very well. This is an expensive way to buy seeds, but is very suitable for those who have difficulty in handling tiny seeds or in bending down to sow them.

Left: using sand to scarify seed. Above left: scarifying using sandpaper. Above right: soaking sweet pea seeds.

composts (substrates)

Please be fussy about your seed compost. Do not use soil taken directly from the garden, or a poor quality compost. Garden soil is not sterile, so your lovingly collected seeds will have to compete with weed seed, which will certainly win. Old compost may contain pests and diseases which will attack and damage your seeds.

You can make your own compost or buy it ready made. There are now some good organic mixes available from garden centres or via the internet.

A good seed compost (substrate) is made up of loam, peat or peat substitute, mixed with sand, grit, bark and a small amount of fertiliser. You can now buy certified organic fertiliser. This is the most convenient choice if you are mixing a small amount of compost as you will get the ratio correct. If you are making a big batch of compost, you can mix your own fertiliser made up of 14g (1/2oz) of ammonium nitrate, 28g (1oz) of potassium nitrate, 56g (2oz) super phosphate, 85g (3oz) chalk, 85g (3oz) magnesium limestone and 14g (1/2oz) of prepared horticultural trace elements. For an ericaceous mix, omit the chalk. This amount of fertiliser should be mixed with 36 litres (8 gallons) of compost. I have taken these measurements from the RHS book on propagation. When I was first taught to make this mix, by a wonderful gardener called Mr. Bell, it was measured in handfuls. One of his handfuls equalled two of mine! However, nowadays we need to be more exact.

Store your compost mixture in a dedicated dustbin which has a close-fitting cover, it will last up to 6 months. Remember, wherever you mix up your compost, make sure it is clean before you start, so you do not mix in old compost, weeds etc. If you have a seed and you are unsure which compost to use, always err on the side of more drainage rather than less. Seeds hate being over wet.

Ingredients

The common ingredients used in compost (substrate) are:

Loam High quality sterilised garden soil. This is very good for seed that has to overwinter outside. When using garden soil in your compost you must sterilise it first. You can do this by baking it in the oven at 200°C (400°F) for 30–40 minutes, or putting it in a plastic bag and placing it in a microwave. In this case, seal and pierce the bag with a few holes. This will stop it popping and plastering your microwave with soil. Cook on high, full power, for 10 minutes.

Peat This is good for seeds that germinate quickly. Do not let it dry out as it can be difficult to re-wet.

Coir Derived from the waste fibre of coconut, coir makes a very good base for soil-less composts (substrate). The seedlings will need feeding within three weeks of germination.

Bark This is a good compost which comes in many grades. You will need the fine- or propagating-grade bark for seed sowing. Bark is particularly useful when making a mix for acid-loving plants.

Clockwise from top left: trays of bark, grit, coir and perlite

Leaf Mould You can use well-rotted, sieved leaf mould as a peat substitute. It is fine for potting on, but it can harbour pests and diseases, so I do not recommend it for seed sowing.

Grit Grit comes in many shapes and sizes and can be added to compost for extra drainage and aeration. It can also be used as a covering for seeds instead of compost.

Sand There are many horticultural sands available, from fine silver sand to very coarse sand. Do not use it directly from the seaside as it will be too salty. It can either be added to compost for extra drainage and aeration, or used as a covering over the seeds instead of compost.

Perlite Perlite is made of sterile, light, expanded volcanic granules and it comes in fine, medium or coarse grades. It can either be added to composts (substrate), for extra drainage and aeration, or used as a covering over the seeds instead of compost.

Vermiculite This is expanded mica, similar to perlite but it holds more water and less air. It comes in various grades, from fine to coarse, and can be added to compost (substrate) for extra drainage and aeration, or used as a covering over the seeds instead of compost.

> **The following mixes are ones I have used in the book**
> **Standard soil-less mix**
> Standard soil-less seed compost (substrate), either peat or peat substitute
> **Standard soil-less mix plus extra sand**
> Standard soil-less seed compost (substrate), either peat or peat substitute mixed with additional silver or fine sand for extra aeration, mixed to a ratio of 3 parts compost + 1 part sand
> **Standard soil-less mix plus extra grit**
> Standard soil-less seed compost (substrate), either peat or peat substitute, mixed with fine (2–3mm/1/8 – 1/4in) grit mixed to a ratio of 3 parts substrate + 1 part fine grit
> **Standard loam mix**
> Standard loam-based seed compost (substrate)
> **Standard loam mix plus extra sand**
> Standard loam-based seed compost (substrate) mixed with coarse horticultural sand to a ratio of 2 parts compost + 1 part sand
> **Standard loam mix plus equal parts sand**
> Standard loam-based seed compost (substrate) mixed with coarse horticultural sand to a ratio of 1 part compost + 1 part sand
> **Standard loam mix plus extra grit**
> Standard loam-based seed compost (substrate) mixed with (5mm / 1/4in) sharp grit to a ratio of 1 part compost + 1 part grit
> **Standard ericaceous (acid mix)**
> Standard ericaceous seed compost (substrate)
> **Standard ericaceous plus extra bark (acid mix)**
> Standard ericaceous seed compost (substrate) mixed with fine-grade composted bark, to a ratio of 2 parts compost + 1 part bark.

equipment

There is a plethora of equipment available for the enthusiastic gardener; however, you do not need to spend a fortune to get started. A few simple and easily obtainable tools are the starting point.

Sieves

If you are collecting your own seed, you will need to clean it.

Sieves of varing mesh sizes are useful for removing the chaff. Be gentle, do not rub the seeds through the mesh as you could damage their outer coating. Shaking out the chaff is much better.

Have dedicated sieves for seed harvesting, do not use them in the kitchen, as some seeds are toxic.

Seed-sowers

The most important implements for sowing seeds are the lines in your hand.

I remember watching well-known gardener, Geoff Hamilton, demonstrating seed sowing many years ago. He said that the lines in our hands evolved so that we could sow seed successfully! How true; you can control the flow of seeds very easily with a bit of practise. For those of you with hot hands, a 3 x 5 card, folded in half, makes an ideal sowing implement. You can see the seed and control the flow by pinching the card. Otherwise, there are numerous seed-sowers on the market. Choose something simple, you do not want to spend hours setting it up.

Seed flats

The seed flat should be 2 in. deep and have adequate drainage holes in the bottom.

The flats should be sturdy enough not to lose their shape when full of wet soil mix, nor to bend when you pick them up or move them. They should be big enough not to dry out too quickly, but not so large that they are awkward to move or lift. I prefer seed flats of 9½ x 14½ in. There are now cell pack inserts available in units of 6 to 200. They are very useful if you only have a small amount of seed to sow. If the seedlings do not like being transplanted, you can pop the entire cell directly into a pot or the flowerbed without disturbing the root ball. The disadvantage of cellpacks is that they dry out much more quickly than seed flats, so they need to be inspected more regularly.

Seed flats and cell packs can be made of plastic, reconstituted peat, or polystyrene. They all have their advantages and disadvantages. You can also buy filled flats; however, I do not recommend these as they can be difficult to 'wet up' prior to sowing.

Pots and containers

There is a wide range of pots and containers available. Plastic pots are light, easy to clean, and retain moisture. Clay pots look good, give better aeration and drainage, but are heavy and difficult to clean.

You can also get plastic sleeves that you fill with soil mix; however, they are awkward to pick up. There are pots made of reconstituted peat or cardboard. These are fine, but they can disintegrate very easily and are no use for wintering outside.

There is a new pot on the market, which looks very exciting. It is made from elephant grass (*Miscanthus* x *giganteus*) bound with natural tree resin. These pots are firm enough for overwintering spring plantings, but they decompose eventually, when planted in the garden.

Labels

The label is essential for identifying what you have sown.

You can have copper labels that last for ever; however, once they are engraved, they are difficult to reuse. There are also black plastic scratch labels that look smart and are permanent, but become brittle with age and are not reusable. White plastic labels are cheap, although they too become brittle with age. You can write on them in pencil, which fades, but does make them reusable.

Waterproof pen

This is very useful for writing on labels, or on plastic bags for storing seeds in the refrigerator.

Plastic bags

I seem to get through endless supplies of plastic bags but, luckily, they are reusable.

I use them for storing seeds in the refrigerator, for collecting cuttings, and also for making a tent or small greenhouse over containers. Once used, wash out carefully, turn inside out, and dry on the washing line.

Watering can and spray

As watering plays such a vital part in the greenhouse and garden, it is worth investing in a good plastic or galvanized watering can with a fine brass rose, which should be turned upwards when watering to create a light spray.

When sowing very fine seeds, you might find a sprayer easier to use. There are two kinds available, the small hand sprayer, which is very useful but does not hold a large amount of water, or a large pump sprayer, which can be a bit heavy when full but does mean you do not have to fill it up continuously when you are sowing a lot of seed.

Top left: equipment for chaffing; top right: seed flats and flowerpots; bottom right: two different seed sowers, labels, marker pen, folded card;
bottom left: watering can, label

Propagator

Using a propagator to start the seeds germinating allows you to adjust the humidity, temperature, air flow, and light to suit specific seeds, giving the seeds the best chance of germination.

There are lots of propagators available. Small ones hold four pots and can be placed on a window-sill to use the natural light. Propagators with a heating element provide the seedlings with bottom heat. They are all useful, and your choice will depend on your needs and your budget. When you buy a propagator, heated or unheated, do make sure that the lid has a vent which allows the excess humidity to be controlled. Otherwise, you can find that some seeds will rot if the humidity gets too high.

For the garden

When sowing directly into the soil in the garden, you will need some basic equipment:

Garden line This can be very basic. I have two stout sticks and a length of twine. You can, of course, buy purpose-made garden lines, which also have a depth gauge marked on them. Their use is to give a good, straight line when establishing a drill prior to sowing.

Measuring dibble A dibble is used for making holes in the soil. It is always useful when sowing large seeds, allowing you to sow each one at the same depth.

Cloches You can buy a wide variety of cloches, ranging from lovely, Victorian, bell-shaped replicas to the expandable cloches that can cover a whole drill. Alternatively, you can cut the bottom off a plastic bottle and put that over a seed or young plant. Protection is essential if you want to start your seeds off in the garden early in the season.

Floating row covers I have found these a wonderful invention, light and reusable. They protect crops from light frosts and act as a barrier against pests, including carrot-root fly and flea beetle.

general instructions on how to sow seeds Fill your clean seed flat, pots or cellpacks with the mixed soil mix, smooth over, tap down and water in well.

Sowing seeds in flats or pots

Before you start sowing your seed, it is a good idea to prepare everything that you need first, just as you would when you cook a meal from a recipe.

large seeds (3–28 seeds per ounce) Space large seeds, which are easy to handle, evenly on the surface of compost, or sow them individually. Then press them gently into the compost to a depth equal to that of the seed.

medium-sized seed (28–2,800 seeds per ounce) is also easy to handle. Space the seeds out evenly on the surface of the compost, then press them in gently until you can no longer see them—just below the surface of the compost.

> **Before you start …**
> Have your seed flats, pots or cell packs clean and ready.
>
> A clean area where you will be putting your seed flats.
>
> Have the labels and a waterproof pen to hand.
>
> Have the soil mix prepared.
>
> Have the seed prepared for sowing.
>
> Have an elastic band or clip ready, to reseal the seed packet if you are using bought seed.

small seeds (2,800–28,000 seeds per ounce) should be tipped in small quantities into the palm of your hand. Allow the seeds to settle in the crease of your palm and then let a small amount trickle onto the surface of the soil mix.

tiny seed (28,000–56,000 seeds per ounce) If the seed is very small, use a label to control the flow of the seed.

minute seed (56,000–140,000 seeds per ounce) A very small amount of fine seed should be put into the crease of a folded 3 x 5 card. To sow the seed, tap the card gently. This will allow you to see what you are sowing and will also enable you to sow thinly on the surface of the soil mix.

extremely fine seed (appears as dust) (over 140,000 seeds per ounce) Mix extremely fine seeds with talcum powder or extrafine white flour, as this will make them easier to see. Put a very small amount of the seed mix into the crease of a folded 3 x 5 card. Tap the card gently to sow the seed thinly on the surface of the soil mix.

Whichever method you use, cover the seeds, following the instructions specific to the plant, using either perlite, vermiculite, horticultural sand or coarse grit. Label with the plant's name and the date. Place the flat or pot under protection, in a cold frame or on a hard surface outdoors. Watch your watering! Keep it to the absolute minimum until germination has taken place, but do not allow the compost to dry out.

Germination

The average germination time is 10–14 days. However, some seeds take over a year.

The most important thing is to watch your watering. For very fine seed, use a sprayer, not a watering can. If you are growing seeds using bottom heat, remove the container from the heat, when 70 percent of the seedlings have emerged, and place in a warm, light, airy place to grow on. Seedlings left on the heat for too long grow soft and leggy and become prone to disease. If you have sown too thickly and all the seedlings emerge, remove a few gently using tweezers. This will allow more room for the other seedlings to develop and encourage air movement, preventing mildew or other root or stem problems associated with overcrowding.

Once the seedlings are large enough to handle, plant them out into a prepared site or pot them on.

Pest control under protection

The major pests of plants grown under protection are aphids whitefly, both of them potentially fatal to young seedlings. To control these pests organically, I suggest the following:

Sticky yellow traps hung in the greenhouse will help you identify which pest you have.

Use horticultural soap, mixed with rainwater, to control aphids, red spider mite and whitefly.

Use the predatory *Encarsia formosa*, a minute parasitic wasp, if you have an infestation of whitefly.

Use *Phytoseiulus persimilis* to control red spider mite.

The above predators can be used only when the night temperature does not fall below 45°F. Do not use a predator and horticultural soap simultaneously.

"Damping-off" is a disease caused by fungi in nonsterile soil or dirty containers. There is no organic remedy, so, before you start, make sure everything is clean and your soil is sterile.

Powdery mildew is a common fungal disease, which can occur when the plants are overcrowded. It is prevented by watering well during dry spells, following the recommended planting distances, and clearing away any fallen leaves.

Sowing seeds in a seed bed

To save space in the main garden, seed can be sown fairly closely in a separate seed bed, from which the young plants will later be lifted and transplanted into their permanent positions.

What went wrong?

If you have sown your seeds according to the instructions and there is no sign of germination:

1 Check how old your seeds are. Most vegetable seeds are best sown fresh or kept for no more than a few years. For example, chervil, angelica, and lovage are only viable for one year.

2 You might have sown too deeply, so that the seed takes longer to emerge.

3 Check that your seed has not rotted in overwet soil.

4 As a final resort, try and find a seed in the soil mix. If it is very small, get a magnifying glass and give it a thorough inspection! You can tell quickly if all is well. The seed should have swelled to at least double its original size, and it should be splitting its casing. If nothing has changed, then I suggest you put it outside on a flat surface and be patient. Nature is marvelous. The seed might germinate a few weeks later or next year.

If your seedlings die unexpectedly:

1 You might have overwatered them.

2 You might have used old soil mix or a dirty container, causing the plants to "damp off".

3 You are stale water and have developed phythium, a waterborne pathogen that causes fungal disease.

Half the flat has germinated and the other half has not, what should you do?

If the existing seedlings are large enough to handle, prick them out carefully and pot up.

If you have used a seed flat or pot, fill the gap created by the removal of the seedlings with fresh soil mix, so that the container is full again. If you have used cell packs, leave the cells empty, so that you know which is which. Make a new label. Write the original date of the sowing on it and add the date of the first potting.

Place the container in a cold frame and restart the whole sequence the following season.

This is quite a common method for growing vegetables, perennials, shrubs, and trees. The advantage is that you have one area to protect from the weather and from pests, instead of different patches throughout the garden. The negative aspect is that the roots can, and do, get damaged when they are lifted. This will check the growth of the young plant. This is not so important for perennials, shrubs, and trees, but it can be a disadvantage when growing vegetables. I recommend a raised- or deep-bed system for growing vegetables, so that the seedlings can remain undisturbed.

Sowing seeds in the open ground

Before sowing seeds in the open ground, prepare the site thoroughly.

The time of year for digging will depend on the crop. I prefer to use a raised- or deep-bed system, with beds 3 ft. wide. The maximum size for beds in my garden is 5 ft., so that I can reach both sides without treading on the soil. Once it is dug thoroughly , feed the soil with well-rotted manure or compost. This can be either your own, homemade, well-rotted compost or farmyard manure. In the very early spring, weather permitting, dig the site over again to make a very fine tilth. As soon as the temperature starts to rise, you can prepare to sow. If the weather has been very dry, water the site well before sowing.

Using the garden line, make a straight drill with the edge of a hoe to the required depth. The drill should be twice the depth of the seed, unless instructed otherwise. Stand on a plank so you do not compact the soil, sow the seeds thinly and evenly along the drill, then cover gently with the soil, being careful not to dislodge the seed. "Water in", using the fine rose of the watering can. Label the row at both ends with the name of the plant and the date. When the seedlings emerge, thin to the recommended spacing. Do not let the seedlings dry out, and protect if necessary from pests and unseasonable weather, using fine netting, floating row covers or cloches.

If you want to "scatter sow" a flower-bed with a splash of colorful annuals, prepare the site well by digging over to remove any weeds, then rake to create a fine, free-draining tilth. Sow the seed thinly to the depth of the seed, cover with soil, water in well and label. When the seedlings emerge, thin to the required spacing. Do not let the seedlings dry out, and protect them if necessary from pests and unseasonable weather, using floating row covers or cloches.

Pest control for seeds grown outside

Young seedlings are a "bonne bouche" for all passing slugs, caterpillars, cutworms, earwigs, wood lice and wrens, so do not thin too enthusiastically.

If, for example, the final spacing is to be 2 in. between each plant, thin initially only to 1 in. As the plants mature, thin down to 2 in. The best control for slugs is a night patrol with a flashlight, rubber gloves, and a bucket of water. Or try scattering salt, charcoal, grit, or eggshell around your plants, or erect a ring of dry holly leaves.

There is now a form of biological control that can be used on slugs. It is a microscopic worm that infects the slugs with a bacterium to stop them feeding within a week and kills them within two. This is an expensive method, but worth it in cases of serious infestation.

The grub of carrot-root fly tunnels into the roots of plants during early summer, especially those of the Apiaceae family which have a long taproot. Feeding the plants with liquid seaweed after transplanting, is one deterrent. Another is to make a polythene barrier 30 in. around the crop during midspring, or to cover the young plants with floating row covers. By sowing seeds after midsummer you might avoid an attack by root flies.

Flea beetle makes little holes in plants leaves, especially those belonging to the Brassicaceae family. The major attacks happen in late spring, especially when the oil seed rape goes into flower. Cover young plants with floating row covers as a physical barrier.

Lunar sowing

Sowing seeds to a lunar calendar is not as strange as it sounds.

If you stop to consider the power that the moon exerts on the ocean tides and inland water, then it is logical to consider this when sowing seed. The time to plant, or transplant, annuals, perennials, trees, leafy vegetables, cucumbers, parsley, peppers, and grain crops, is from the appearance of the new moon, and the period during which it is growing (or waxing). Once the third quarter is reached, when the moon is on the wane, it is the time to plant root crops. It is said that when the moon is in the last quarter, one should not plant anything at all. This is thought to be the best time to weed or cultivate, ready for the first quarter to come again. If you want more information on this growing method, there are a number of specialty books available (see 'Further Reading').

Remember ...

Remember to label and date, not only when you collect and packet the seeds, but also when you have sown them.

Do not overwater—seedlings watered too much can rot. Even if they do survive, they will become very weak and transplant badly.

If you have used bottom heat to germinate your seeds, remove the container from the heat as soon as 70 percent of the crop has germinated.

If you leave young seedlings on the heat for too long, they will become weak and stretched and you will have difficulty in transplanting them.

Remember to put your remaining seeds away into a dark, cool. dry place after sowing.

Do not sow the complete contents of the seed packet in one go, sow half at a time as an insurance against crop failure.

Do not sow too many varieties of seed at once, as they will all have to be potted up or planted out at the same time.

Do not harvest too much seed, as you might not have the space to dry it.

Remember to sow thinly. Overcrowding causes "damping off".

acknowledgements

I am very grateful to my fellow exhibitors on the RHS Flower Show circuit. My special thanks goes to Dave Clarke for his knowledge on grasses, Brian Goodey for his knowledge on cacti, Brian Hiley for his knowledge on tender perennials. A very big thank you to Jim Juby from DT Browns for his immense patience and infinite knowledge of vegetables.

I thank Marianne Majerus for her beautiful photographs, Vanessa Courtier for designing the book, Helena Attlee for wading through my script and Helen Woodhall for keeping everything together. A special thank you goes to Kyle for producing another beautiful book and to Anthea for being a tame rottweiler. And finally this would not have been possible without the support of my family Mac, Hannah, Alistair and my animals, William, Blackie and Catmint.